257
Topics in Current Chemistry

Topics in Current Chemistry

Recently Published and Forthcoming Volumes

Prebiotic Chemistry
Volume Editor: Walde, P.
Vol. 259, 2005

Supramolecular Dye Chemistry
Volume Editor: Wuerthner, F.
Vol. 258, 2005

Molecular Wires
From Design to Properties
Volume Editor: De Cola, L.
Vol. 257, 2005

Low Molecular Mass Gelators
Design, Self-Assembly, Function
Volume Editor: Fages, F.
Vol. 256, 2005

Anion Sensing
Volume Editor: Stibor, I.
Vol. 255, 2005

Organic Solid State Reactions
Volume Editor: Toda, F.
Vol. 254, 2005

DNA Binders and Related Subjects
Volume Editors:Waring, M.J., Chaires, J.B.
Vol. 253, 2005

Contrast Agents III
Volume Editor: Krause,W.
Vol. 252, 2005

Chalcogenocarboxylic Acid Derivatives
Volume Editor: Kato, S.
Vol. 251, 2005

New Aspects in Phosphorus Chemistry V
Volume Editor:Majoral, J.-P.
Vol. 250, 2005

Templates in Chemistry II
Volume Editors: Schalley, C.A.,Vögtle, F., Dötz, K.H.
Vol. 249, 2005

Templates in Chemistry I
Volume Editors: Schalley, C.A.,Vögtle, F., Dötz, K.H.
Vol. 248, 2004

Collagen
Volume Editors: Brinckmann, J., Notbohm, H., Müller, P.K.
Vol. 247, 2005

New Techniques in Solid-State NMR
Volume Editor: Klinowski, J.
Vol. 246, 2005

Functional Molecular Nanostructures
Volume Editor: Schlüter,A.D.
Vol. 245, 2005

Natural Product Synthesis II
Volume Editor:Mulzer, J.
Vol. 244, 2005

Natural Product Synthesis I
Volume Editor:Mulzer, J.
Vol. 243, 2005

Immobilized Catalysts
Volume Editor: Kirschning, A.
Vol. 242, 2004

Transition Metal and Rare Earth
Compounds III
Volume Editor: Yersin, H.
Vol. 241, 2004

The Chemistry of Pheromones and Other Semiochemicals II
Volume Editor: Schulz, S.
Vol. 240, 2005

The Chemistry of Pheromones and Other Semiochemicals I
Volume Editor: Schulz, S.
Vol. 239, 2004

Orotidine Monophosphate Decarboxylase
Volume Editors: Lee, J.K., Tantillo,D.J.
Vol. 238, 2004

Molecular Wires

From Design to Properties

Volume Editor: Luisa De Cola

With contributions by
C. Chiorboli · G. H. Gelinck · F. C. Grozema · M. T. Indelli ·
D. K. James · M. A. Ratner · F. Scandola · L. D. A. Siebbeles ·
J. M. Tour · J. M. Warman · M. R. Wasielewski · E. A. Weiss ·
K. M.-C. Wong · V. W.-W. Yam

Springer

The series *Topics in Current Chemistry* presents critical reviews of the present and future trends in modern chemical research. The scope of coverage includes all areas of chemical science including the interfaces with related disciplines such as biology, medicine and materials science. The goal of each thematic volume is to give the nonspecialist reader, whether at the university or in industry, a comprehensive overview of an area where new insights are emerging that are of interest to a larger scientific audience.
As a rule, contributions are specially commissioned. The editors and publishers will, however, always be pleased to receive suggestions and supplementary information. Papers are accepted for *Topics in Current Chemistry* in English.
In references *Topics in Current Chemistry* is abbreviated Top Curr Chem and is cited as a journal.

Visit the TCC content at springerlink.com

ISSN 0340-1022
ISBN-10 3-642-43932-2 **Springer Berlin Heidelberg New York**
ISBN-13 978-3-642-43932-2 **Springer Berlin Heidelberg New York**
DOI 10.1007/b98356

Springer is a part of Springer Science+Business Media
springeronline.com

Softcover re-print of the Hardcover 1st edition 2005

Cover design: KünkelLopka GmbH, Heidelberg; *Design & Production* GmbH, Heidelberg
Typesetting and Production: LE-TEX Jelonek, Schmidt & Vöckler GbR, Leipzig

Printed on acid-free paper 02/3141xv – 5 4 3 2 1 0

Volume Editor

Professor Luisa De Cola
Westfälische Wilhelms-Universität
Physikalisches Institut
Mendelstr. 7
D-48149 Münster,
Germany
decola@uni-muenster.de

Editorial Board

Topics in Current Chemistry also Available Electronically

For all customers who have a standing order to Topics in Current Chemistry, we offer the electronic version via SpringerLink free of charge. Please contact your librarian who can receive a password for free access to the full articles by registration at:

springerlink.com

If you do not have a subscription, you can still view the tables of contents of the volumes and the abstract of each article by going to the SpringerLink Homepage,clicking on "Browse by Online Libraries", then "Chemical Sciences", and finally choose Topics in Current Chemistry.

You will find information about the

- Editorial Board
- Aims and Scope
- Instructions for Authors
- Sample Contribution

at springeronline.com using the search function.

Preface

In the emerging field of nanoscience the desire to construct objects which in the molecular scale can resemble and act as macroscopic devices is a constant trend. Amongst the simplest but also most attractive functional component *molecular wires* are surely the most investigated. The reason for such interest is quite obvious since wires, in today world, are used all around us and are the most indispensable unit to assemble any type of electronic devices. But at this point the question is what is a molecular wire? Amongst the several definitions the one which will be addressed in this issue and which in my opinion best describes the discrete molecular systems dealt with in this issue is: a molecular wire is a molecule or an assembly of molecules able to strongly electronically couple the terminal sites in order to mediate energy and charge transport over long distance. Since many years several scientist have attempted to synthesize unidimensional molecular systems able to vectorially transport energy or charge from one terminal site to the other one. Often the terminal units have been donor and acceptor groups, and light has been used as energy source to promote the energy or charge migration. More recently a lot of attention has focused on the possibility to interface molecular structures between electrodes of conductive materials, enabling a direct measurement of the *conductivity* of the system. Despite the differences in experimental set-up and the conditions applied, it is clear that there is a good agreement on the insulating or conductive properties of the molecules obtained by the various methods. However, in the construction of a multicomponent system great attention must be paid on the terminal units since the electronic coupling between these terminal groups and what is in between can dramatically influence the overall behavior.

The misuse of the words *molecular wire* has lead to confusion in the definition and many rod-like systems, just because their linear and rigid structure, have been labeled as wires. I would like to stress that in many cases the lost in the transport is dramatic and an exponential decrease of the conductivity is observed with increasing distance between the terminal sites resulting in a few nanometer conductance which cannot be addressed as a wired behavior.

I believe that effort in connecting, with molecular wires, components possessing different properties and address them with different inputs can lead to a much more interesting way to conceive nanoelectronics which goes far behind than just trying to interconnect macro and nanoworld. The possibility to have a complete electronic or photo-addressable circuit which requires no external *physical* interface to be operational is just a vision at the moment but the realization of components self-consistent and molecularly connected is reality.

In this issue the synthesis and characterization of molecular wires based on organic or metal complexes components, their properties and some theoretical background will be illustrated. It is clear that we are very far for a full overview but I hope that the readers will find the field as much exciting as it is for me.

Münster, 15 July 2005 Luisa De Cola

Contents

Luminescent Molecular Rods – Transition-Metal Alkynyl Complexes
V. W.-W. Yam · K. M.-C. Wong 1

Molecular Wires
D. K. James · J. M. Tour 33

Photoinduced Electron/Energy Transfer Across Molecular Bridges in Binuclear Metal Complexes
C. Chiorboli · M. T. Indelli · F. Scandola 63

Molecules as Wires: Molecule-Assisted Movement of Charge and Energy
E. A. Weiss · M. R. Wasielewski · M. A. Ratner 103

The Opto-Electronic Properties of Isolated Phenylenevinylene Molecular Wires
F. C. Grozema · L. D. A. Siebbeles · G. H. Gelinck · J. M. Warman 135

Author Index Volumes 251–257 165

Subject Index 169

Contents of Volume 249

Templates in Chemistry II

Volume Editors: Schalley, Christoph A.; Vögtle, Fritz; Dötz, Karl H.
ISBN: 3-540-23087-4

First Considerations: Principles, Classification, and History
D.H. Busch

Macrocycle Synthesis Through Templation
Z.R. Laughrey · B.C. Gibb

Macrocycles and Complex Three-Dimensional Structures Comprising Pt(II) Building Blocks
A. Kaiser · P. Bäuerle

Templated Synthesis of Interlocked Molecules
F. Aricó · J. D. Badjic · S.J. Cantrill · A.H. Flood · K.C.F. Leung · Y. Liu · J.F. Stoddart

Molecular Knots
C. Dietrich-Buchecker · B.X. Colasson · J.-P. Sauvage

Templation in Noncovalent Synthesis of Hydrogen-Bonded Rosettes
M. Crego-Calama · D.N. Reinhoudt · M.G.J. ten Cate

Imprinted Polymers
A.J. Hall · M. Emgenbroich · B. Sellergren

Top Curr Chem (2005) 257: 1–32
DOI 10.1007/b136069

Published online: 6 July 2005

Luminescent Molecular Rods – Transition-Metal Alkynyl Complexes

Vivian Wing-Wah Yam (✉) · Keith Man-Chung Wong

Centre for Carbon-Rich Molecular and Nano-Scale Metal-Based Materials Research, and Department of Chemistry, The University of Hong Kong, Pokfulam Road, Hong Kong, P.R. China
wwyam@hku.hk

1	**Introduction**	2
2	**Rhenium(I) and Ruthenium(II)**	4
2.1	Rhenium(I) Polypyridines	4
2.2	Ruthenium(II) Polypyridines	6
3	**Platinum(II)**	8
3.1	Platinum(II) Phosphines	8
3.2	Platinum(II) Polypyridines	12
4	**Copper(I) and Silver(I)**	18
5	**Gold(I)**	21
6	**Mixed-Metal**	24
6.1	Rhenium(I)-Copper(I)/Silver(I)	25
6.2	Rhenium(I)-Iron(II)	26
6.3	Gold(I)-Rhenium(I)	27
7	**Concluding Remarks**	27
References		28

Abstract A number of transition-metal complexes have been reported to exhibit rich luminescence, usually originating from phosphorescence. Such luminescence properties of the triplet excited state with a large Stoke's shift, long lifetime, high luminescence quantum yield as well as lower excitation energy, are envisaged to serve as an ideal candidate in the area of potential applications for chemosensors, dye-sensitized solar cells, flat panel displays, optics, new materials and biological sciences. Organic alkynes (poly-ynes), with extended or conjugated π-systems and rigid structure with linear geometry, have become a significant research area due to their novel electronic and physical properties and their potential applications in nanotechnology. Owing to the presence of unsaturated *sp*-hybridized carbon atoms, the alkynyl unit can serve as a versatile building block in the construction of alkynyl transition-metal complexes, not only through σ-bonding but also via π-bonding interactions. By incorporation of linear alkynyl groups into luminescent transition-metal complexes, the alkynyl moiety with good σ-donor, π-donor and π-acceptor abilities is envisaged to tune or perturb the emission behaviors, including emission energy (color), intensity and lifetime by its role as an auxiliary ligand as well as to govern the emission origin from its direct involvement. This review summarizes recent efforts on the synthesis of luminescent rod-like alkynyl complexes with different classes of transition metals and details the effects of the introduction of alkynyl groups on the luminescence properties of the complexes.

Keywords Luminescence · Alkynyl · Metal–metal interaction · Phosphorescence

Abbreviations

^{t}Bu$_2$bpy	4,4′-di-*tert*-butyl-2,2′-bipyridine
Bpy	2,2′-bipyridine
Fc	ferrocene
Np	naphthylene
Me$_2$bpy	4,4′-dimethyl-2,2′-bipyridine
(CF$_3$)$_2$bpy	4,4′-bis(trifluoromethyl)-2,2′-bipyridine
Phen	1,10-phenanthroline
Dppm	bis(diphenylphosphino)methane
Dppa	*N*,*N*-bis(diphenylphosphino)amine
HOMO	highest occupied molecular orbital
LUMO	lowest unoccupied molecular orbital

1 Introduction

Luminescent transition-metal complexes, in particular those with triplet excited states that show emission of a phosphorescence nature arising from the enhanced intersystem crossing as a result of the larger spin-orbit coupling due to the presence of the heavy metal center, have attracted considerable attention in recent years. In addition to studies that provide the fundamental information for the elucidation of electronic structures and examination of photoinduced electron or energy transfer processes [1–4], extensive studies on such luminescent complexes have also led to their potential applications in the area of chemosensing, dye-sensitized solar cells, flat panel displays, organic light-emitting devices (OLEDs), optical materials and biological applications. For example, a number of ruthenium(II) complexes have been utilized as dyes, which were immobilized on a TiO_2 surface via anchoring groups. Photoexcitation of the metal-to-ligand charge transfer (MLCT) leads to injection of electrons into the conduction band of the TiO_2 for the conversion of solar energy into electricity [5–8]. Due to the triplet phosphorescence nature, a number of luminescent transition-metal complexes have been employed as the emitting layer for OLED fabrication with improvements in the electroluminescence (EL) intensity, quantum efficiency as well as the tuning of EL color [9–14].

Organic alkynyl compounds (poly-ynes) with extended π-conjugated systems, rigid structure and linear geometry have attached enormous attention due to their novel electronic and physical properties as well as their ability to serve as potential candidates for development in nanotechnology [15–17]. In view of this, incorporation of alkynyl groups into transition-metal complexes may lead to systems with unique optical, electronic and physical properties that may find potential applications as non-linear optical materials, liquid crystals, molecular electronics and wires [18–25]. Owing to the presence of unsaturated *sp*-hybridized C-atoms, the alkynyl group can serve as a versatile building block in the construction of alkynyl transition-metal complexes in the σ-bonding and π-bonding modes. By introduction of the linear alkynyl group with the unique properties of having good σ-donor, π-acceptor and π-donor abilities into luminescent transition-metal complexes, fine-tuning and perturbation of emission behaviors, in the sense of energy (color), emission intensity and lifetime could be envisaged through their role as an auxiliary ligand. In some cases, the alkynyl groups may exert their influence in perturbing the emission properties by their direct involvement in the emission origin. The introduction of heavy metal atoms into the alkynyl backbone in metal-containing oligo- and poly-ynes also provided a versatile method to improve the chances of producing triplet emitters, leading to long-lived triplet emission, as a result of a larger spin-orbit coupling constant and an enhanced intersystem crossing efficiency.

In several classes of transition-metal complexes, the presence of a non-emissive low-lying *d–d* state would quench the luminescence excited state via thermal equilibration or electron/energy transfer [26,27]. Coupling of a strong σ-donating alkynyl ligand to such a system is anticipated to raise the *d–d* state, resulting in improvement or enhancement of the luminescence by increasing the population of the luminescence excited state. On the other hand, the *sp*-hybridized unsaturated $C\equiv C$ group is capable of interacting with other metal centers, for example copper(I), silver(I) or gold(I), through π-bonding fashion [22,28–32]. Therefore, introduction of such metal centers into the rigid rod-like metal alkynyl complex would alter the structural and luminescence properties. Elongation of the alkynyl group has also been reported to vary the linearity and electronic communications of the metal alkynyl complexes [25,33]. It is envisaged that such an extension of the conjugated π-system would lead to perturbation of the luminescence properties.

2
Rhenium(I) and Ruthenium(II)

2.1
Rhenium(I) Polypyridines

In 1974, Wrighton and coworkers reported the photoluminescence studies of a new class of rhenium(I) tricarbonyl diimine complexes, $[Re(CO)_3(N-N)Cl]$ $(N-N$ = phen or related diimine ligand), exhibiting octahedral geometry with the three carbonyl groups arranged in a *facial* fashion [34]. The emission origin has been assigned to be derived from the triplet $d\pi(Re) \rightarrow \pi^*(N-N)$ metal-to-ligand charge transfer (MLCT) excited state. Since then, numerous studies on the luminescence properties of related complexes, including derivatives of the chlororhenium(I) tricarbonyl diimine complexes such as those obtained by the replacement of the chloro ligand with pyridine, phosphine, nitrile or isonitrile ligands, have been reported [35–40]. In 1995, Yam and coworkers reported the first synthesis and photophysical properties of luminescent rhenium(I) complexes incorporating alkynyl ligands [41]. The treatment of the precursor complex, $[Re(CO)_3(^tBu_2bpy)Cl]$ with a lithiated alkynyl reagent, prepared in situ from the reaction of ^{n}BuLi with the corresponding terminal alkyne, afforded the desired alkynylrhenium(I) products, $[Re(CO)_3(^tBu_2bpy)(C\equiv C-R)]$ (R = alkyl or aryl; Scheme 1). Later on, a more general synthetic protocol was developed to enable the synthesis of a variety of luminescent alkynylrhenium(I) complexes with different diimine as well as diynyl ligands, with the general formula, $[Re(CO)_3(N-N)(C\equiv C)_n-R]$ (N – N = bpy, $^tBu_2bpy$, Me_2bpy, $(CF_3)_2bpy$, phen; n = 1 or 2) (Scheme 1) [42–44]. Such a pathway has also opened up the possibility of employing the trimethylsilyl-protected alkynes instead of the terminal alkynes and extended the versatility and flexibility of the synthetic

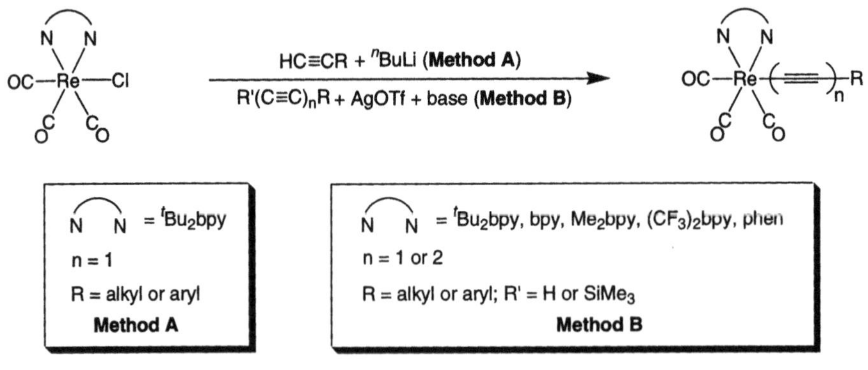

Scheme 1 Synthetic routes of rhenium(I) monoynyl an diynyl complexes

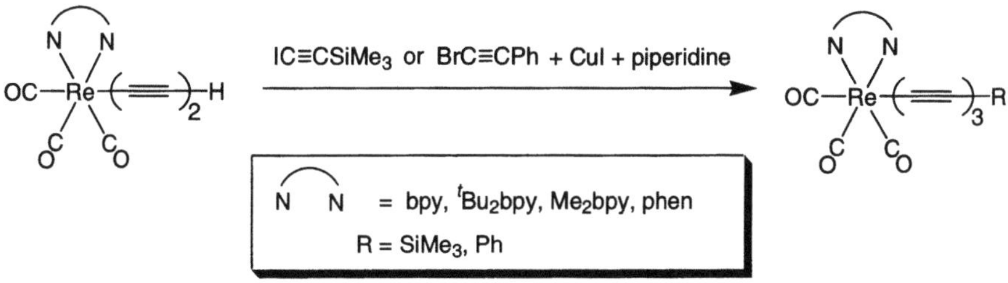

Scheme 2 Synthetic routes of rhenium(I) triynyl complexes

methodology. Depending on the nature of the diimine and alkynyl ligands, either reaction could be employed to give the desired target complexes. Extension of the work to the tri-ynyl system has been achieved by the reaction of the butadiynyl complex, $[Re(CO)_3(N-N)(C\equiv C-C\equiv CH)]$, containing a terminal acetylenic proton, with an excess of the bromo- or iodo-alkyne reagent in the presence of copper(I) catalyst (Scheme 2) [44, 45]. In principle, elongation to higher order poly-ynyl complexes is feasible by the utilization of the above methodology in a series of stepwise reactions.

Upon excitation at $\lambda > 350$ nm in the solid state and in fluid solution, these rhenium(I) alkynyls exhibited intense orange phosphorescence, which generally occurs at lower energy compared to the chloro precursor, $[Re(CO)_3(N-N)Cl]$, attributed to the presence of the strong σ-donating alkynyl group. The luminescence energies have been found to be dependent on the nature of both the diimine and alkynyl ligands. The origin of such luminescence has been ascribed to a ^{3}MLCT $[d\pi(Re) \rightarrow \pi^*(N-N)]$ excited state, probably with some mixing of $[\pi(C\equiv C) \rightarrow \pi^*(N-N)]$ ligand-to-ligand charge transfer (LLCT) character taking into consideration the relatively high-lying filled π $(C\equiv C)$ orbital.

For the complexes with the same alkynyl ligand, higher luminescence energy was observed for the complexes containing better electron-donating substituents on the diimine ligand, resulting in the order of ^{3}MLCT energy: **1** > **6** > **7** (Fig. 1) [43–46]. On the other hand, **2**, with a strongly electron-donating alkynyl group showed lower luminescence energy than **1**, and even much lower than **3**, in line with the energy level of the $d\pi(Re)$ orbital. It is noteworthy that the longer the carbon chain of the alkynyl group is, the higher is the luminescence energy, i.e. **5** > **4** > **1**. Despite the fact that a better energy

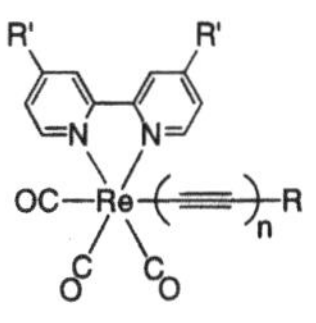

R' = ^{t}Bu		
n = 1	n = 2	n = 3
R = C_6H_5 **1**	R = C_6H_5 **4**	R = C_6H_5 **5**
= $^nC_6H_{17}$ **2**		
= H **3**		

R' = H	R' = CF_3
n = 1	n = 1
R = C_6H_5 **6**	R = C_6H_5 **7**

Fig. 1 Mononuclear alkynylrhenium(I) complexes of **1**–**7**

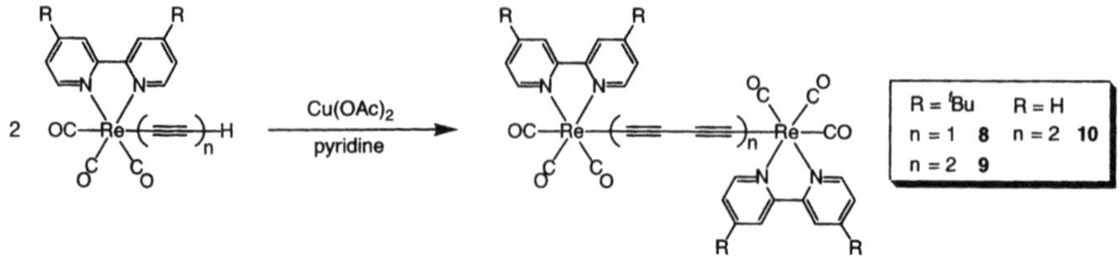

Scheme 3 Synthetic routes of dinuclear alkynylrhenium(I) complexes **8–10**

match with the $d\pi$(Re) orbital in the filled-filled $p\pi - d\pi$ overlap would be expected from the higher-lying $\pi(C \equiv C)$ orbital of the longer alkynyl analogue, a poorer overlap integral actually would be resulted due to delocalization of the electron density across the alkynyl chain, as supported by molecular orbital calculations.

The synthesis of dinuclear luminescent rhenium(I) complexes tethered by a linear rod-like oligo-ynyl bridge has been accomplished by oxidative homo-coupling reaction in the presence of copper(II) catalyst (Scheme 3) [43, 44, 47, 48]. Owing to the absence of *tert*-butyl substituents on the bpy ligand, **10** is not soluble in common organic solvents. Complexes **8–10** showed intense luminescence, typical of ^{3}MLCT emission with some mixing of LLCT character. In the solid state at 77 K, the phosphorescence origins of **8** and **9** have been assigned as derived from the metal perturbed intra-ligand (IL) $[\pi(C \equiv C) \rightarrow \pi^*(C \equiv C)]$ excited state, which has been supported by the observation of a structured emission band with vibrational progressional spacings of ca. 2100 cm^{-1}, characteristic of the $\nu(C \equiv C)$ stretch.

2.2 Ruthenium(II) Polypyridines

The polypyridyl ruthenium(II) system, similar to the rhenium(I) diimine complexes, is another class of the most studied luminescent transition-metal complexes. While most studies on ruthenium alkynyl complexes were focused on the use of phosphine ligands, corresponding studies on nitrogen donor ligands are relatively rare. As an extension of the luminescence studies on alkynylrhenium(I), Yam and coworkers attempted to design and prepare the luminescent alkynylruthenium(II) complexes in 1998. Reaction of *cis*-$[Ru(N-N)_2(Me_2CO)_2]^{2+}$ with terminal alkyne yielded an unprecedented ruthenium(II) (aryl)-carbonyl complex, *cis*-$[Ru(N-N)_2(CO)(\eta-CH_2-C_6H_4-R-4)]^+$. It was proposed that the terminal alkyne first coordinated to the ruthenium(II) metal center to form a vinylidene intermediate, with the electron-deficient α-carbon highly susceptible to nucleophilic attack. Finally, a ruthenium(II) (aryl)-carbonyl complex was obtained by the nucleophilic attack of a water molecule on the α-carbon of the vinylidene ligand via the hydroxycarbene complex intermediate

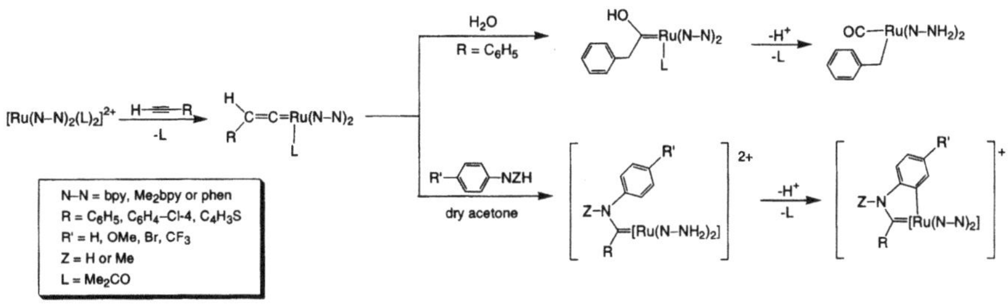

Scheme 4 Reactions of *cis*-$[Ru(N-N)_2(Me_2CO)_2]^{2+}$ with terminal alkynes in the presence of water or amines

(Scheme 4) [49–51]. On the other hand, if a nucleophilic primary or secondary aniline was present in dry acetone, an orthometallated ruthenium(II) aminocarbene complex, *cis*-$[Ru(N-N)_2\{=C(CH_2-R)NZC_6H_3-R'\}]^+$, was formed instead via the formation of an aminocarbene complex and subsequent orthometallation of the phenyl ring (Scheme 4) [49–51]. All these aminocarbene complexes exhibited intense emission in various media and such emission origin was assigned to $d\pi(Ru) \rightarrow \pi^*(N-N)$ ^{3}MLCT excited state. A change in the substituents on the aminocarbene ligands has been shown to have relatively little influence on the emission energies of these complexes with the same diimine ligand. The emission energy was found to be sensitive to the nature of the diimine ligand, as reflected by a relatively significant blue shift in emission energies from the complex with the bpy ligand to another with the Me_2bpy ligand. The presence of electron-donating methyl substituents on the bpy ligand would render the π^* orbital of the diimine ligand higher-lying in energy and thus resulting in a higher MLCT emission energy. A semi-empirical MO theoretical calculation showed that the HOMO was predominantly ruthenium(II) metal center character while the LUMO was mainly localized on the diimine ligands. Such results are consistent with the ^{3}MLCT assignment. In addition, the respective binuclear and crown-ether containing ruthenium(II) aminocarbene complexes have also been obtained from the reaction with functionalized alkynyl ligands and their luminescence behavior as well as the cation binding properties studied [51].

More recently, Che and coworkers reported the synthesis and photophysical properties of a series of polypyridyl ruthenium(II) alkynyl complexes, $[Ru(trpy)(bpy)(C\equiv C-R)]^+$ (Scheme 5) [52], formed by the substitution reaction of $[Ru(trpy)(bpy)(H_2O)]^{2+}$ with a deprotonated terminal alkyne reagent. Excitation of these complexes resulted in red emission in both the solid state and in acetonitrile solution. Typical of other polypyridyl ruthenium(II) systems, the emission was assigned as $d\pi(Ru) \rightarrow \pi^*$(polypyridyl) ^{3}MLCT character. Since the alkynyl ligand incorporating a stronger σ-donating substituent would destabilize the $d\pi(Ru)$ energy level to a larger

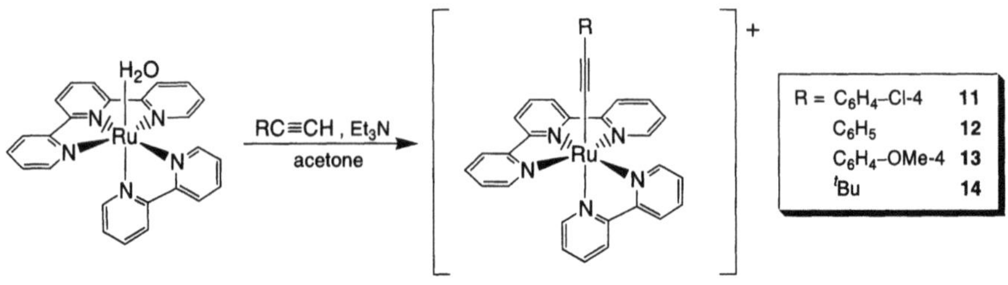

Scheme 5 Synthetic routes of polypyridyl ruthenium(II) alkynyl complexes **11–14**

extent, the emission energy decreases in the order: **11** > **12** > **13** > **14**, in line with the MLCT assignment.

3 Platinum(II)

3.1 Platinum(II) Phosphines

Alkynylplatinum(II) phosphine complexes, containing two phosphine ligands and two alkynyl groups with square-planar geometry, exist in *cis*- or *trans*-configuration depending on the phosphine type and reaction conditions. A number of monomeric, oligomeric and polymeric platinum(II) compounds in *trans*-configuration, in which the platinum(II) metal center and the two alkynyl ligands adopt a linear geometry, have been reported in view of their ideal conformations as potential candidates in the construction of molecular rods or wires. The general synthetic method involves the dehydrohalogenation of the *cis*- or *trans*-$[Pt(PR_3)_2Cl_2]$ starting material with the alkynyl ligand in the presence of copper(I) catalyst and amine base (Scheme 6) [24, 53–68]. The luminescence studies of

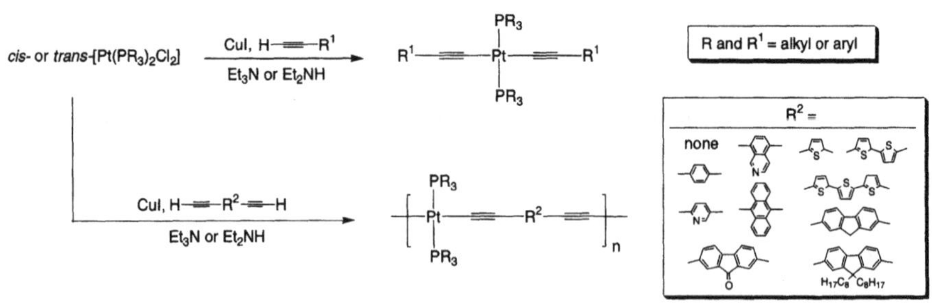

Scheme 6 Synthetic routes of alkynylplatinum(II) phosphine complexes

trans-[Pt(PR_3)$_2$(C $\equiv$ CH)$_2$] and *trans*-[Pt(PEt_3)$_2$(C $\equiv$ CPh)$_2$] were reported by DeGraff, Lukehart, Demas and coworkers [53], with an excited state assignment of Pt $\rightarrow \pi^*$(C $\equiv$ C) MLCT in character, which has been further supported by resonance Raman spectroscopy [54]. On the other hand, the luminescence properties of a series of platinum(II) alkynyl polymers, *trans*-[Pt(PR_3)$_2$(C $\equiv$ C – R – C $\equiv$ C)$_2$]$_n$ (R = none or aromatic spacer), have been extensively studied by Lewis, Marder, Friend, Raithby and coworkers [55–64], and the origin of such luminescence has been ascribed to be derived from a metal perturbed intra-ligand π–π^*(C $\equiv$ C) excited state. By an iterative-convergent approach with trimethylsilyl protected alkynyl groups, Schanze and coworkers recently reported the synthesis of a class of monodisperse platinum(II) alkynyl oligomers (Fig. 2) [69]. The complexes, **16** and **17** were found to exhibit luminescent $^1[\pi$–$\pi^*]$ and $^3[\pi$–$\pi^*]$ states of the conjugated system, and the effect of oligomer length on the luminescence properties has been elucidated. The maxima of the absorption and fluorescence bands shifted to red from **15** to **19**, but the difference between **19** and **20** is small, indicating that the effective conjugation length in the singlet state (*S*1) is about 6 repeat units. Gladysz and coworkers have prepared a series of binuclear alkynylplatinum(II) complexes, *trans*-[(C_6F_5)Pt{μ– $Ph_2P(CH_2)_mPPh_2$}$_2$(C $\equiv$ C)$_n$Pt(C_6F_5)] (m = 10, 11, 14 and 18; n = 4 and 6) [70] and *trans*-[{Pt(PR_3)$_2$(C_6F_5)}(C $\equiv$ C)$_n${Pt(PR_3)$_2$(C_6F_5)$_2$}] [71] (R = Et or *p*-tol; n = 1, 3, 4, 6, 8, 10 and 12) while the {Pt}(C $\equiv$ C)$_{12}${Pt} is the longest metal poly-ynyl compound to date (Fig. 3). UV-Vis spectroscopic studies on *trans*-[{Pt(PR_3)$_2$(C_6F_5)}(C $\equiv$ C)$_n${Pt(PR_3)$_2$(C_6F_5)}] showed that the most intense band shifted monotonically to lower energy upon increasing the number of C $\equiv$ C units.

Shaw, Pringle and coworkers reported the synthesis of a dinuclear platinum(II) alkynyl complex, [Pt_2(dppm)$_2$(C $\equiv$ CPh)$_4$], showing an interesting double rod-like structure, with two *trans*-[Pt(C $\equiv$ CPh)$_2$] moieties tethered by two diphosphine ligands [72]. A number of dinuclear d^8–d^8 metal complexes with well-defined metal–metal distances have been reported, and their unique luminescence features associated with the presence of metal–metal interactions have been extensively studied [73–78]. In order to probe the role of metal–metal interactions in determining the luminescence properties, the design and synthesis of discrete molecules with known metal–metal distances are required. The complex, [Pt_2(dppm)$_2$(C $\equiv$ CPh)$_4$] **21**,

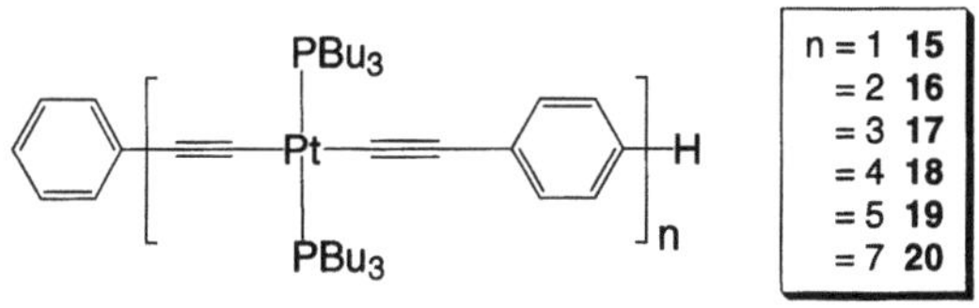

Fig. 2 A class of monodisperse platinum(II) alkynyl oligomers of 15–20

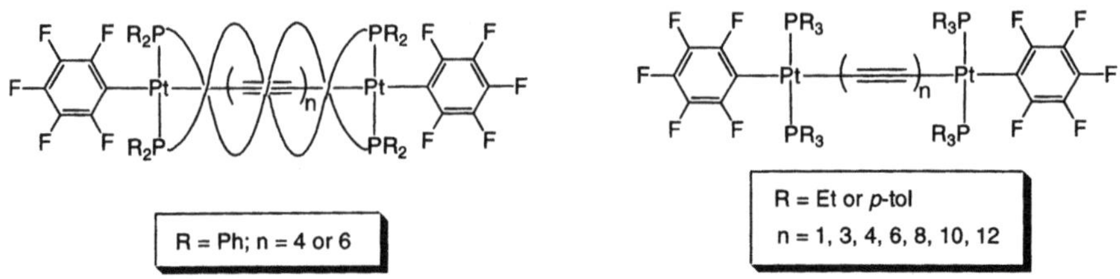

Fig. 3 Dinuclear alkynyl-bridged platinum(II) phosphine complexes

was found to show luminescence in different media by Yam and coworkers, and its analogues **22–24** with different alkynyl groups were also prepared (Scheme 7) [79–81]. The luminescence properties and origins have been elucidated by systematic variation of the nature of the ligands and the structural features. By comparison with the mononuclear counterpart, *trans*-[Pt(dppm – P)$_2$(C ≡ CPh)$_2$], in which the luminescence was attributed to an admixture of Pt → π^*(C ≡ C)MLCT/π → π^*(C ≡ C) IL excited states, the observation of a lower emission energy in **21** was suggested to be as a result of the involvement of metal–metal interaction. The luminescence origins of **21–24** were assigned as derived from the triplet metal-metal-to-ligand charge transfer (^{3}MMLCT) [d_σ^*(Pt$_2$) → p_σ(Pt$_2$)/π^*(C ≡ CR)] state, where d_σ^* and p_σ denote the antibonding combination of d_z^2(Pt) –d_z^2(Pt) interaction and bonding combination of p_z(Pt) –p_z(Pt) interaction, respectively, taking the Pt – Pt direction as the *z*-axis. The crystal structures of **21–24** provided the information of the Pt – Pt distances in order of decreasing length: **21** $\approx$ **22** > **24** > **23**. The ^{3}MMLCT energy were found to be dependent on the nature of alkynyl groups as well as the Pt – Pt distance. In general, complexes with the less electron-rich alkynyl ligands would have a lower-lying π^*(C ≡ CR) orbital level and hence a smaller HOMO-LUMO energy gap, giving rise to lower emission energy. In addition, complexes with shorter Pt – Pt distance would also result in lower emission energy due to the higher-lying d_σ^*(Pt$_2$) (HOMO) energy level and lower-lying p_σ(Pt$_2$) orbital energy derived from the stronger Pt – Pt interaction. In view of these observations, **23**, with the least electron-rich alkynyl ligand and the shortest Pt – Pt distance, gave the lowest emission energy amongst the series.

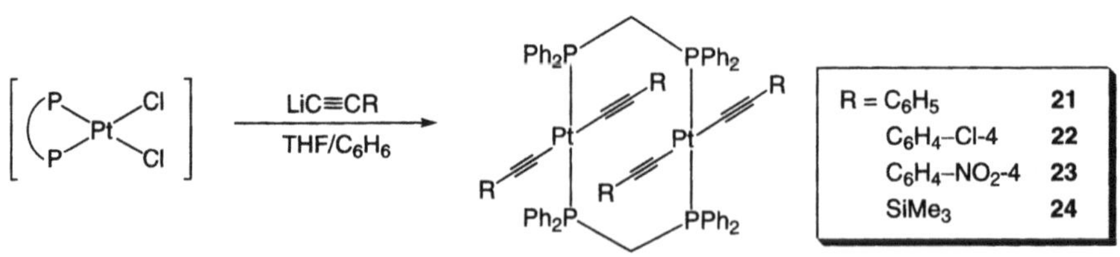

Scheme 7 Synthetic routes of face-to-face dinuclear alkynylplatinum(II) complexes **21–24**

The unique structure of **21** provides preorganized sites for the ready encapsulation of other metal centers through π-interaction with the alkynyl group which serves as η^2-ligands to form tweezer-like complexes **25** and **26** (Scheme 8) [79–81]. Perturbation of the emission properties has been accomplished by a change in the Pt – Pt distance as well as the π-acceptor ability of the alkynyl unit via copper(I) or silver(I) coordination. Lower MMLCT emission energy was found in **25** and **26** relative to that in **21**. The observation of shorter Pt – Pt distances in **25** and **26**, compared with their precursor **21**, was ascribed to the coordination of copper(I) or silver(I) metal center which pulled the platinum atoms into close proximity as a result of the reduced donor strength of the alkynyl ligands as well as the steric requirements. Upon encapsulation of such Lewis acids, the shortening of the Pt – Pt distance would raise the $d\sigma^*(Pt_2)$ orbital energy. Meanwhile, a lowering of the $p\sigma(Pt_2)/\pi^*(C \equiv CPh)$ orbital energy (LUMO) would occur as a result of an increase in the alkynyl π-accepting ability. Consequently, lower emission energies were observed in **25** and **26**, resulting from a narrowing of the HOMO-LUMO energy gap. Slightly higher emission energy of **26** relative to **25** was observed due to the weaker Lewis acidity of Ag(I) as well as its larger ionic radius compared to Cu(I), resulting in a higher energy MMLCT transition, in addition to the slightly longer Pt – Pt distance in **26**.

Alternatively, perturbation of the luminescence behavior has also been achieved by introduction of an extra attachment site in the alkynyl group for further coordination. By using the "metalloligand" approach, assembly of four platinum(II) terpyridyl units, $[Pt(trpy)]^{2+}$, into the pyridyl anchors in **27** afforded a novel hexanuclear platinum(II) complex, $[Pt_2(dppm)_2\{C \equiv C - C_5H_4N - Pt(trpy)\}_4]^{4+}$ **28** (Scheme 9) [82]. Similar to the case of the π-encapsulated complexes **25** and **26** versus **21**, a shorter Pt – Pt distance and lower emission energy were observed in **28**, relative to **27**. Employing the same arguments, coordination of four electron-deficient plat-

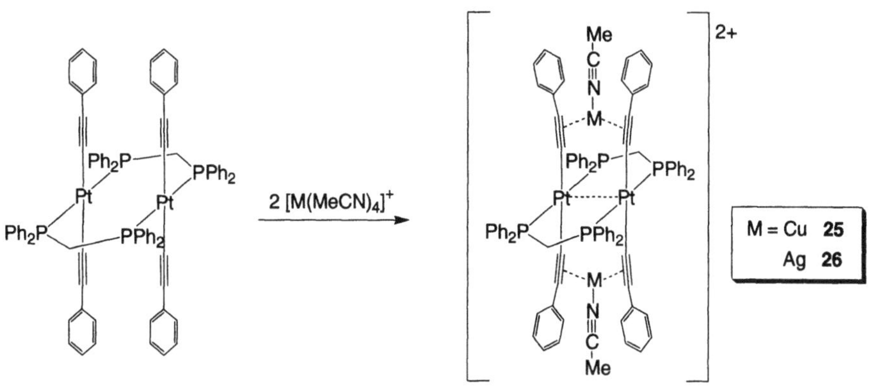

Scheme 8 Synthetic routes of tetranuclear platinum(II) complexes **25** and **26**

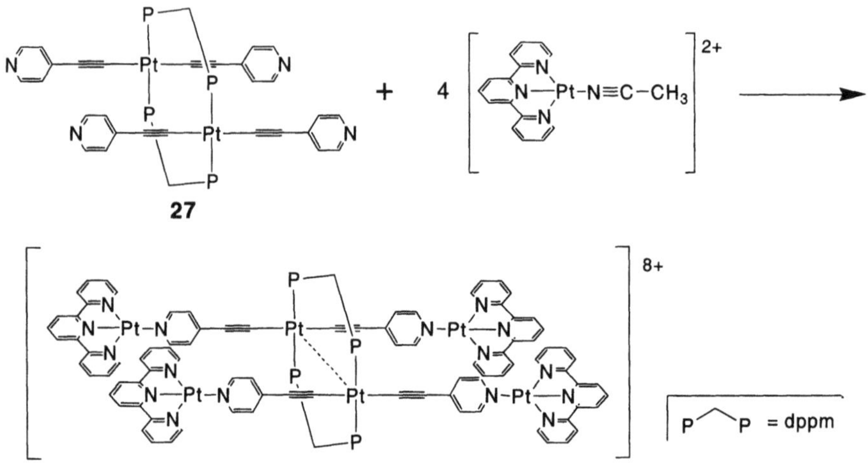

Scheme 9 Synthetic route of hexanuclear platinum(II) complex **28**

inum(II) terpyridyl moieties resulted in the lowering of $p_\sigma(Pt_2)/\pi^*(C\equiv CPh)$ (LUMO) level and shortening of the Pt – Pt distance which in turn would boost the energy level of $d^*_\sigma(Pt_2)$ (HOMO). Both effects would cause a narrowing of the HOMO-LUMO energy gap for this MMLCT transition and account for the red shift in emission energy of **28**.

3.2 Platinum(II) Polypyridines

Apart from the platinum(II) phosphine systems, platinum(II) polypyridyl complexes have also aroused a growing interest due to their intriguing spectroscopic and luminescence behaviors associated with the propensity to form Pt – Pt and π-π interactions. Examples include the platinum(II) terpyridyl complexes, $[Pt(N-N-N)L]^{n+}$ ($n = 1$ or 2; N – N – N = terpyridyl ligand; L = anionic or neutral ligand), which have been extensively studied and shown to exhibit rich luminescence attributed to $d\pi(Pt) \rightarrow \pi^*(N-N-N)$ ^{3}MLCT origin [27, 83–96]. In some cases, such ^{3}MLCT excited states would be quenched by the presence of a low-lying non-emissive *d*–*d* ligand field (LF) or ligand-to-ligand charge transfer (LLCT) states, depending on the nature of the auxiliary ligand, *L* and the terpyridyl chelator [27, 96–98]. Yam and coworkers first reported the preparation of a luminescent alkynylplatinum(II) terpyridyl system $[Pt(N-N-N)(C\equiv C-R)]^+$ [93] by employing a synthetic methodology similar to that for rhenium(I) alkynyl complexes, in which the labile acetonitrile ligand of the precursor complex $[Pt(N-N-N)(MeCN)]^{2+}$ was substituted by the alkynyl group (Scheme 10). The luminescence exhibited by these complexes was attributed to the $d\pi(Pt) \rightarrow \pi^*(N-N-N)$ ^{3}MLCT

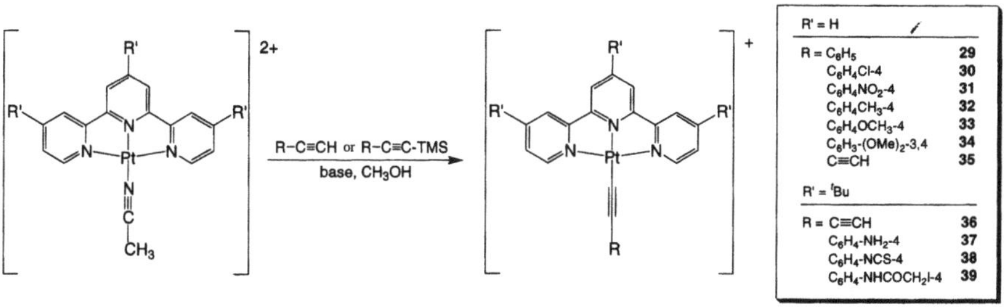

Scheme 10 Synthetic routes of alkynylplatinum(II) terpyridyl complexes **29–39**

excited state, with some mixing of $\pi(C\equiv C)\rightarrow\pi^*(N-N-N)$ ^{3}LLCT character. Compared to the chloro-analogue, $[Pt(N-N-N)Cl]^+$, lower luminescence energy was observed in the platinum(II) alkynyl complexes due to the presence of the higher-lying $d\pi$(Pt) orbital resulting from coordination of the strong σ-donating alkynyl group. In general, the stronger the electron-donating ability of the alkynyl ligand is, the lower is the luminescence energy observed. This is in line with the fact that better electron-donating alkynyl ligands would destabilize the $d\pi$(Pt) orbital to a larger extent. The order of the luminescence energy: **31**> **30** > **29** > **32**, is also in line with the MLCT/LLCT assignment. Higher ^{3}MLCT emission energy has also been observed in **36** relative to its analogue **35**, as a result of a higher-lying $\pi^*(N-N-N)$ orbital in **36** due to the presence of three electron-donating *tert*-butyl substituents on the terpyridyl ligand. It is interesting to note that **33**, **34** and **37** are non-emissive in fluid solution and in the solid state. Such phenomenon may be ascribed to quenching of the emissive ^{3}MLCT excited state by photoinduced electron transfer (PET), in which the electron is transferred from the electron-rich substituent on the alkynyl ligand to the platinum metal center. Alternatively, the presence of an energetically accessible or lower-lying ^{3}LLCT excited state, due to the effect of the electron-donating substituent, may cause quenching of the ^{3}MLCT emission.

Through the modification of the amino group on **37**, complexes **38** and **39** containing the isothiocyanate and iodoacetamide functional groups, were designed for luminescence bio-labeling from the ready reaction with the primary amine and sulfhydryl group, respectively, of the biomolecules [96]. Human serum albumin (HSA) has been labeled with **38** and **39** to afford the corresponding bioconjugates, **38-HSA** and **39-HSA** (Scheme 11), which are also luminescent in the visible region upon photoexcitation. The luminescence properties of the bio-conjugates **38-HSA** and **39-HSA**, together with their parent labels have been investigated and elucidated. The bioconjugate **38-HSA** resulting from the isothiocyanate complex **38** was found to emit with a different luminescence color when compared to its parent label, while the bioconjugate **39-HSA** derived from the iodoacetamide complex **39** was found to emit with an enhanced luminescence intensity.

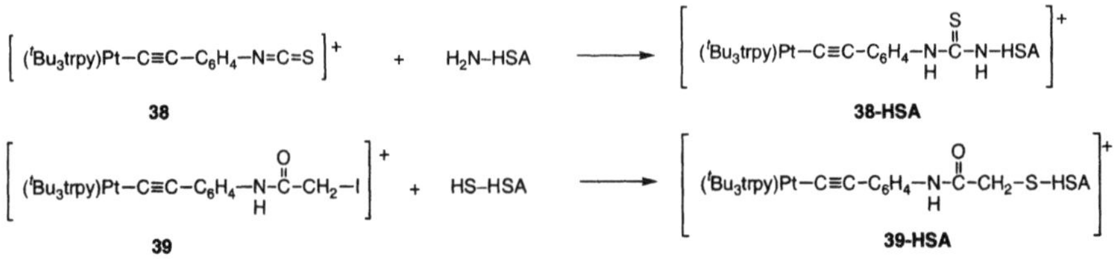

Scheme 11 Bio-labeling of HSA with **38** and **39** to afford the corresponding bioconjugates **38-HSA** and **39-HSA**

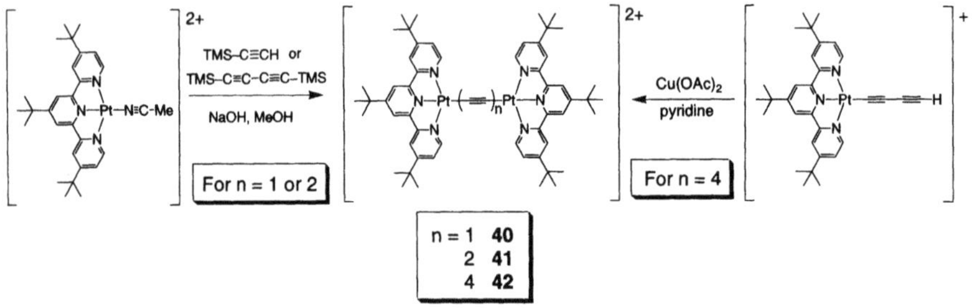

Scheme 12 Synthetic routes of dinuclear alkynylplatinum(II) terpyridyl complexes **40–42**

A series of luminescent alkynyl-bridged dinuclear complexes with platinum(II) terpyridyl units serving as end-capped termini, $[Pt(^tBu_3trpy)(C\equiv C)_nPt(^tBu_3trpy)]^{2+}$ ($n = 1, 2$ or 4), was designed and synthesized (Scheme 12) [94]. Complexes **40** and **41** were directly obtained by the reaction of $[Pt(^tBu_3trpy)(MeCN)]^{2+}$ with the corresponding alkynyl ligands, while **42** was prepared via an oxidative homo-coupling reaction from **36**. The luminescence behavior together with the structure-property relationships, in particular those related to the effect of the alkynyl chain length, were studied. In order to minimize the influence of Pt – Pt or π–π interactions on the luminescence behavior, the ligand $^tBu_3trpy$ with three bulky substituents was selected. The emission bands observed in **40** and **41** were assigned as derived from ^{3}MLCT $d\pi(Pt) \rightarrow \pi^*(^tBu_3trpy)$ origin, with some mixing of ^{3}LLCT $\pi(C\equiv C) \rightarrow \pi^*(^tBu_3trpy)$ character. On the other hand, the emission band of **42** was rich in vibronic structures with vibrational progressional spacings of about 2100 cm^{-1}, typical of the $\nu(C\equiv C)$ stretch in the ground state. This observation indicates the increasing contribution of the alkynyl bridge in the excited state properties and the assignment of an emission origin to the predominantly ^{3}IL $\pi \rightarrow \pi^*\{(C\equiv C)_4\}$ excited state was made. Higher emission energy was observed in **41** compared to **40** and such a luminescence shift to higher energy with increasing number of $C\equiv C$ units is in contrast to the common observation of the red shift in transition energy with increasing the $C\equiv C$ chain length in organic polyynes and in other metal alkynyl systems. Similar observation has been found in the rhenium(I) alkynyl complexes, and the same argument could also be applied to this system. The blue shift in luminescence observed on going from **40** to **41** is likely due to the greater stabilization of the $d\pi$ orbital from the filled-filled interaction between $p\pi$–$d\pi$ overlap. Despite a higher $d\pi$(Pt) orbital would be expected due to the better energy match between the $d\pi$(Pt) and the π(C ≡ C) orbitals upon increasing the length of the alkynyl chain, an opposing effect from the decreased overlap integral attributed to the better delocalization of electron density across the C ≡ C unit would lead to a lower energy level of $d\pi$(Pt) orbital. As a result, the luminescence energy of **40** occurs at higher energy relative to **41**.

Square-planar platinum(II) polypyridyl complexes are well known to exhibit interesting polymorphism in the solid state, which has been attributed to the variation of Pt – Pt distances or degrees of π–π interaction. Different photophysical and luminescence properties are also observed for different polymorphs and these are found to be dependent on such changes in the Pt – Pt or π–π distances. For instance, in the complex [Pt(bpy)Cl_2], two crystal forms exist and the Pt – Pt distances for the yellow and red forms are found to be 4.44 and 3.45 Å, respectively, from their crystal packing [99, 100]. Interestingly, complex **35** was found to exist in two crystal forms, the dark green and red forms and both of which have been structurally characterized [95]. Although the molecular structures were the same in both forms, different crystal packing arrangements have been revealed. The dark green form exists as a linear chain with the platinum atoms equally spaced with short intermolecular Pt – Pt distances of 3.388 Å, while the red form exhibits a zigzag arrangement to give dimeric structures with alternating Pt – Pt distances of 3.394 and 3.648 Å (Fig. 4). Dissolution of both forms in acetone or acetonitrile gave the same absorption spectrum, indicating that the same molecular species exists in the solution state. There was a dramatic color change

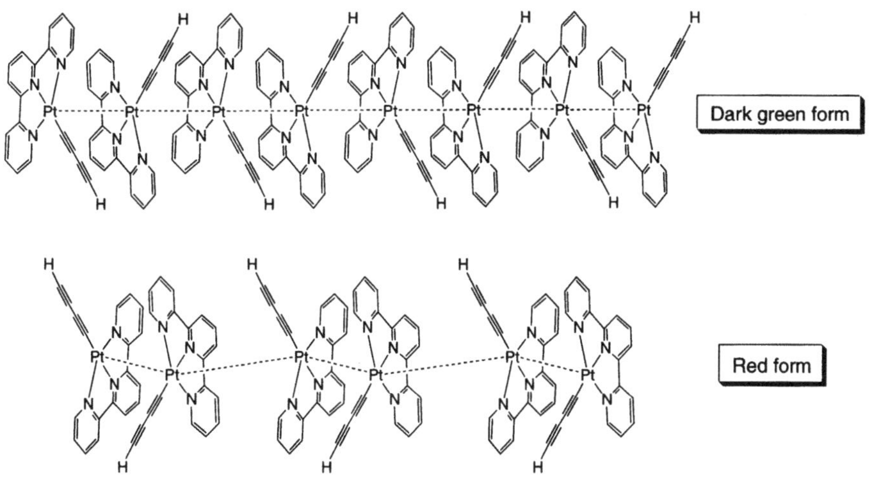

Fig. 4 Crystal packing diagrams of **35** in dark green form and red form

with a tremendous emission enhancement of **35** in solution upon varying the composition of the solvent. Upon increasing the diethyl ether content in solution, the color of the solution changed dramatically from yellow to green to blue, as reflected from its electronic absorption spectra with a growth of the band at 615 nm and a concomitant drop in intensity of the band at 416 nm. Such drastic color changes displayed were ascribed to the formation of aggregates as a result of increasing the non-solvent content of diethyl ether. The new absorption band at 615 nm was assigned to the MMLCT transition of the aggregated species; the formation of which was facilitated by the propensity of this class of compounds to form Pt – Pt interactions and π-π stacking. Apart from the spectral changes, a tremendous emission enhancement at 785 nm was also induced by the change of solvent composition. In view of the close resemblance of the emission intensity variation at 785 nm to the absorbance changes at 615 nm as a function of diethyl ether content, the emission band at 785 nm and the absorption band at 615 nm were assigned to be from the same MMLCT transition origin.

Attempts to prepare a pentanuclear Pt(II) complex by the assembly of homoleptic $[Pt(C\equiv C-C_5H_4N)_4]^{2-}$ with $[Pt(^tBu_3trpy)(MeCN)]^{2+}$, instead afforded an unprecedented dinuclear complex **43** where two $[Pt(^tBu_3trpy)]$ units were connected by an ethynylpyridine ligand (Scheme 13) [101]. Intense luminescence was exhibited both in the solid state and in fluid solution and the origin was assigned as derived from the ^{3}MLCT $d\pi(Pt) \rightarrow \pi^*(^tBu_3trpy)$ excited state, mixed with some ^{3}LLCT $\pi(C\equiv C) \rightarrow \pi^*(^tBu_3trpy)$ character. In view of the higher-lying orbital energy of the platinum atom attached to the alkynyl group, as compared to the other platinum atom adjacent to the pyridyl unit, the lowest energy excited state was assigned as due to the alkynylplatinum(II) terpyridyl moiety, given the better electron-donating ability and the anionic nature of the alkynyl ligand.

A related tridentate cyclometalated platinum(II) alkynyl system, $[Pt(C-N-N)(C\equiv C-R)]$, was prepared by employing Sonogashira's conditions by Che and coworkers (Scheme 14) [14, 102] and luminescence studies have been described. These complexes exhibited intense luminescence of different colors and origins, depending on the nature of the cylometalated and alkynyl ligands, as well as the relative platinum(II) orbital energies. By exten-

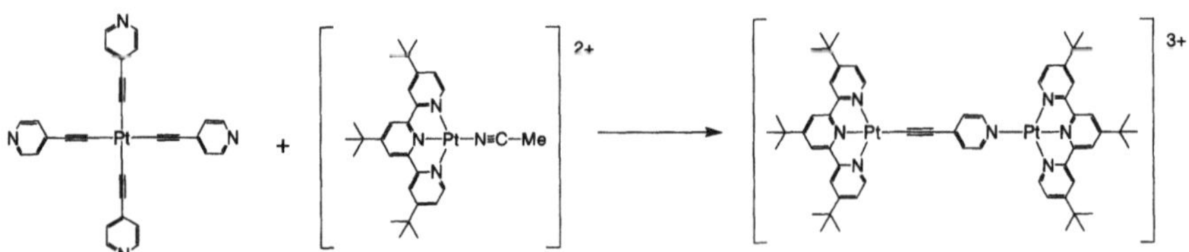

43

Scheme 13 Synthetic route of dinuclear ethynylpridine-bridged platinum(II) complex **43**

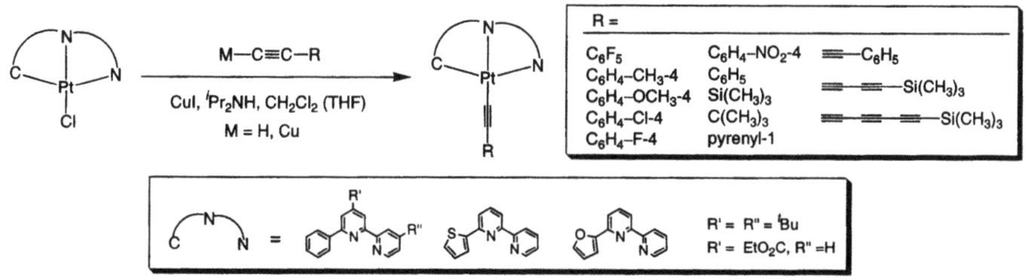

Scheme 14 Synthetic routes of cyclometalated platinum(II) alkynyl complexes

sive modifications of both the tridentate and alkynyl ligands, their structural, photophysical and luminescence properties have been fine-tuned.

The complexes **44–49** are selected as examples for illustrating the possibility of color-tuning from blue to red luminescence with different origins (Fig. 5). In principle, the luminescence origins of the complexes with 6-phenyl-2,2′-bipyridine and simple alkyl or aryl alkynyl ligands were assigned to be derived from the ^{3}MLCT $d\pi(\text{Pt}) \rightarrow \pi^*(\text{C}-\text{N}-\text{N})$ excited state, with some exceptions of intraligand $^3[\pi \rightarrow \pi^*(\text{C}\equiv\text{C}-\text{R})]$ and/or $^3[\pi \rightarrow \pi^*(\text{C}-\text{N}-\text{N})]$ origins in the examples containing substituted 6-aryl-2,2′-bipyridines, electron-deficient and polyaromatic alkynyl ligands, and poly-ynyl groups. The tri-ynyl complex **44** was found to show higher ^{3}MLCT energy compared to the mono-ynyl analogue **45**, and such a blue shift in emission energy upon extending the alkynyl chain length has also been observed in the related platinum(II) terpyridyl [94] and rhenium(I) diimine systems [44, 45]. However, elongating the alkynyl chain to the tetra-ynyl unit in **46** would alter the emission origin from ^{3}MLCT to intraligand $^3[\pi \rightarrow \pi^*\{(\text{C}\equiv\text{C})_4\}]$ excited state, which is believed to be of lower energy than that of the ^{3}MLCT state. The observation of vibronically structured narrow bandwidth emission, similar to the case of $[\text{Pt}(^t\text{Bu}_3\text{trpy})(\text{C}\equiv\text{C})_4\text{Pt}(^t\text{Bu}_3\text{-trpy})]^{2+}$ [94], was suggested to be IL localized

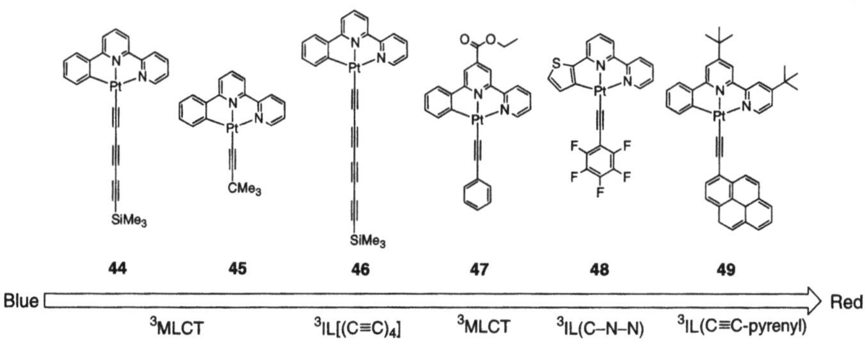

Fig. 5 Cyclometalated platinum(II) alkynyl complexes of **45–49**

$^3[\pi \rightarrow \pi^*]$ emission. A red shift in emission energy of **47** with an ester-substituted C – N – N ligand relative to the unsubstituted analogues **44** and **45** was observed. Such a red shift in emission energy was ascribed to the presence of the electron-withdrawing ester group on the C – N – N ligand, resulting in the lower π^*(C – N – N) orbital energy and hence lower energy ^{3}MLCT emission. An emission of $^3[\pi \rightarrow \pi^*(\mathrm{C-N-N})]$ origin was assigned for **48** with the 6-(2-thienyl-2,2′-bipyridine) ligand based on the fact that the emission energy was insensitive to the substituents on the ethynylbenzene motif and a lower emission energy was observed than their related analogues with 6-phenyl-2,2′-bipyridine. Substantial changes in the nature of the emission origin has also been achieved in **49** by introduction of 1-ethynylpyrene to the [Pt(C – N – N)] motif, in which the intraligand $^3[\pi \rightarrow \pi^*(\mathrm{pyrene})]$ excited state contributed predominantly to the luminescence origin.

In view of the ready tunability of their emission color, these cyclometalated platinum(II) complexes have been utilized as luminescent dopant materials in the fabrication of organic light-emitting devices (OLED), producing electroluminescence (EL) of green to orange-red color with high luminance and efficiency [14, 102].

4 Copper(I) and Silver(I)

The *sp*-unsaturated alkynyl group exhibits a variety of bonding modes with metal centers, of which the most common type in copper(I) and silver(I) is in a $\mu_3-\eta^1$ fashion. Two trinuclear copper(I) alkynyl complexes, $[\mathrm{Cu_3}(\mu-\mathrm{dppm})_3(\mu_3-\eta^1-\mathrm{C{\equiv}C-Ph})]^{2+}$ and $[\mathrm{Cu_3}(\mu-\mathrm{dppm})_3(\mu_3-\eta^1-\mathrm{C{\equiv}C-Ph})_2]^{+}$, were first synthesized and characterized by Gimeno and coworkers [103, 104]. Both have been structurally characterized, in which the three copper(I) atoms were bridged by three dppm ligands with one (the former) or two (the latter) alkynyl groups capping the triangular array of metal atoms in a linear $\mu_3-\eta^1$ fashion. The luminescence studies were reported by Yam and coworkers [43, 105–112] and modification of these two classes of complexes employing different diphosphine ligands and alkynyl groups have been made to elucidate the origin of the luminescence properties. By altering the reaction stoichiometry of the alkyne, the mono-capped complexes were synthesized using a lithiated reagent for deprotonation of the acetylenic proton in THF, while the bi-capped complexes were prepared with sodium hydroxide in THF/MeOH mixture (Scheme 15). All showed intense luminescence both in the solid state and in fluid solution. In general, the more electron-rich the alkynyl ligand is, the lower is the emission energy in both the mono-capped and bi-capped series. In view of this, together with the observation of the short Cu – Cu distances found in their crystal

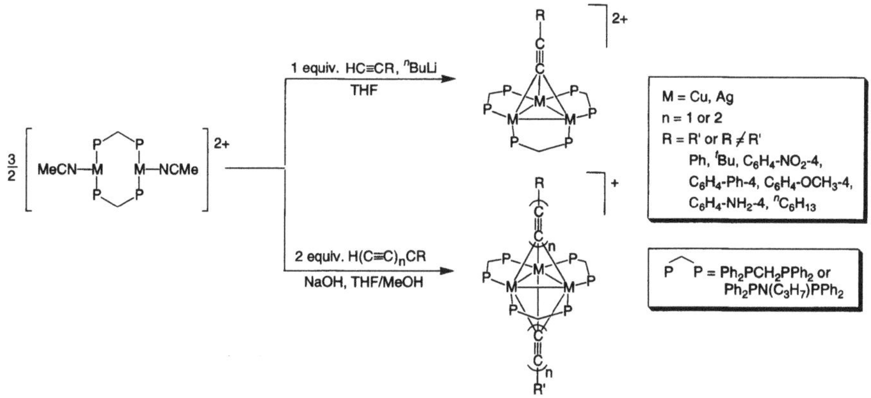

Scheme 15 Synthetic routes of trinuclear copper(I) and silver(I) alkynyl complexes

structures, the emission origin was assigned as derived from an excited state of substantial ligand-to-metal charge transfer (LMCT) $[(C\equiv C-R)\rightarrow Cu_3]$ character, mixed with some metal-centered $[3d^9 4s^1]$ character. Lower emission energy was observed for the mono-capped complexes, relative to the bi-capped system, which was ascribed to the greater stabilization of copper(I) metal centers from the effect of higher overall positive charge, supportive of the assignment of the LMCT state. However, involvement of an intraligand $^3[\pi\rightarrow\pi^*(C\equiv C-R)]$ character in the emission origin was suggested for the complexes containing electron-withdrawing substituents on the alkynyl ligand, such as 4-nitrophenylacetylide or biphenylacetylide, from the fact that the emission bands were very similar to that of the uncoordinated free alkyne. For the di-ynyl complexes **52–55** (Fig. 6), the observation of a structured emission band with vibrational progressional spacings of 1800–2200 cm^{-1}, typical of $\nu(C\equiv C)$ stretch in the ground state, also suggested the involvement of the di-ynyl ligands in the excited state.

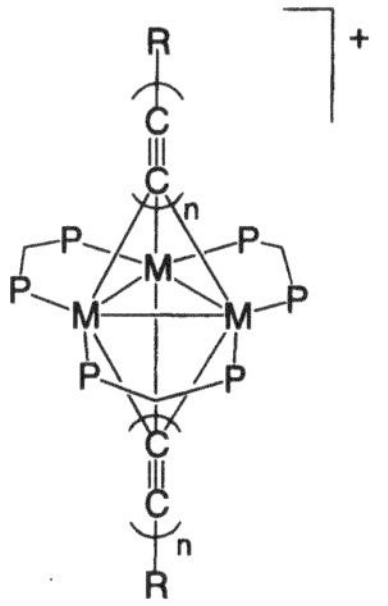

M = Cu			
n = 1		n = 2	
R = H	**50**	R = H	**52**
C_6H_5	**51**	C_6H_5	**53**
		C_6H_4-Me-4	**54**
		C_6H_4-OMe-4	**55**
		$^nC_6H_{13}$	**56**

M = Ag	
n = 2	
R = C_6H_5	**57**
C_6H_4-OMe-4	**58**

P⌒P = dppm

Fig. 6 Trinuclear copper(I) and silver(I) alkynyl complexes of **50–58**

Since copper(I) and silver(I) metal centers are closely related and show similar bonding modes, comparison of the luminescence properties of copper(I) and silver(I) congeners of such complexes would provide a better understanding of the nature of the luminescence origin. Using the same synthetic strategies for the preparation of the copper(I) complexes, simply by replacing $[\{Cu(P-P)(MeCN)\}_2]^{2+}$ with $[\{Ag(P-P)(MeCN)\}_2]^{2+}$ (P – P = diphosphine bridging ligand) as the starting material, luminescent trinuclear mono-capped and bi-capped silver(I) alkynyl complexes with similar structures as the corresponding copper(I) complexes were obtained (Scheme 15) [112–114]. A comparison of the luminescence behaviors of the copper(I) alkynyl complexes with the corresponding silver(I) alkynyl analogues has been made. By keeping the bridging phosphine and alkynyl ligands the same, a higher energy emission was generally observed for the silver(I) analogues. For example, the silver(I) complexes **57** and **58** exhibited higher emission energy than that of the corresponding copper(I) congeners **53** and **55**. Since Cu(I) is much more easily oxidized, given the higher energy of the Cu(I) 3*d* orbitals relative to those of the Ag(I) 4*d* orbitals, the emission assignment of a $[d(Cu/Ag) \rightarrow \pi^*(C\equiv C-C\equiv CR)]$ ^{3}MLCT excited state is in line with the observation of a red shift in emission energy from silver(I) to copper(I) analogues. However, such an assignment of the emissive state as derived from a MLCT transition is less likely, given the relatively small energy difference for the emission of silver(I) complexes with the corresponding copper(I) analogues, from the fact that the ionization energy of Ag^+(g) is nearly 10^4 cm^{-1} larger than that for Cu^+(g) [115]. In view of this, together with the close resemblance of the observed energy shift with similar systems of LMCT origin, the origin of the emission has been proposed to involve substantial ^{3}LMCT $[(C\equiv C-C\equiv CR) \rightarrow Cu_3/Ag_3]$ character, mixed with some metal-centered character. Two hexanuclear complexes, **59** and **60**, were prepared by using 1,4-diethynylbenzene as the bridging ligand (Scheme 16) [113]. The two complexes are isostructural and show a dumb-bell-like structure which consists of two triangular units of copper(I) or silver(I) atoms with three bridging dppm ligands connected by the 1,4-diethynylbenzene moiety. Intense luminescence was observed, which was attributed to originate from the ^{3}LMCT $[(C\equiv C-R) \rightarrow Cu_3]$ state, mixed with some metal-centered $[3d^9 4s^1]$ and IL

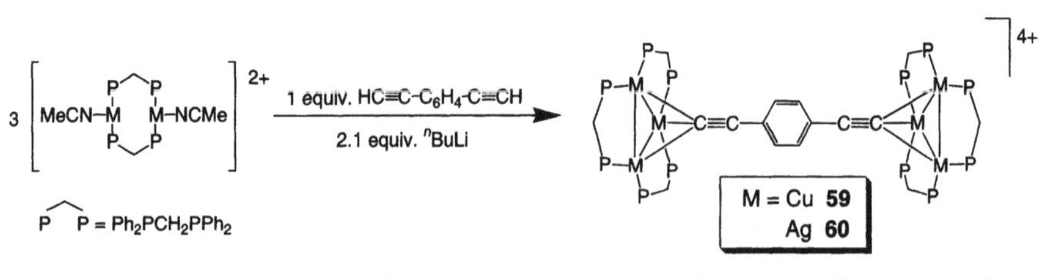

Scheme 16 Synthetic routes of hexanuclear copper(I) and silver(I) alkynyl complexes **59** and **60**

$[\pi \rightarrow \pi^*(C \equiv C)]$ character. Similarly, the silver(I) analogue **60** emitted at higher energy than the copper(I) counterpart **59**.

In order to gain more insights into their electronic structures and the bonding mode between the copper triangle $[Cu_3]$ with the alkynyl groups, theoretical calculations on the hydrogen-substituted model complexes, **50-H–53-H** by replacing the phenyl group of dppm with hydrogen in the mono-ynyl and di-ynyl complexes **50–53** respectively (Fig. 6), have been made [112]. Quantum chemical calculations using density functional theory (DFT) gave satisfactory agreement between the computed bond distances of **51-H–53-H** in the DFT-optimized structures and the respective experimental structural data of **51–53**. From the results of the DFT calculations, the HOMO region was found to consist of four nearly degenerate MOs, while two nearly degenerate MOs were located in the LUMO. For the di-ynyl model **52-H**, two of the HOMOs were predominantly localized on the carbon C4 chain with a lesser extent on the copper(I) and the other two HOMOs were mainly located in the copper(I) triangle motif, while the LUMOs were of π-type carbon in character with contribution from the copper orbitals. Such results are supportive of the assignment of an emission origin to one that has a substantial LMCT parentage, mixed with a metal-centered $nd^9(n+1)s^1$ state as well as intraligand $[\pi \rightarrow \pi^*(C \equiv C - C \equiv CR)]$ character, in particular for the di-ynyl complexes. A larger DFT HOMO-LUMO energy gap was computed for the mono-ynyl model **50-H**, in which their HOMOs were similar in energy and nature to that of **52-H**. On the other hand, the LUMOs, which were weighed on the dHpm (hydrogen-substituted dppm) with less contribution from the carbon fragment and the metal triangle, were calculated to be higher energy than the LUMOs of **52-H**. This is in line with the fact that mono-ynyl complexes exhibited higher emission energy, compared to the di-ynyl analogues. Calculations on different diynyl complex models **53-H–55-H** indicated some significant shift to higher energy of the HOMOs relative to the hydrogenated counterpart **52-H**. Accordingly, these compounds were more readily ionized which is in accordance with the electrochemical properties reported for the diynyl series. The substituted aryl complexes **53–56** were also found to emit at lower energy than that of **52** in the experimental observation. However, one should be aware that the assignments of electronic transitions between metal and/or ligand localized orbitals are only rough approximations because of the possible extensive orbital mixing in these complexes.

5
Gold(I)

Gold(I) complexes may exhibit a wide range of molecular structures with a coordination number of two, three and four. One of the most common coordination geometry amongst them is the two-coordinate linear geometry. The luminescence studies of gold(I) complexes are of particular interest due to the possible observation of Au – Au aurophilic interaction, which would perturb the luminescence behaviors [116–128]. Three classes of two-coordinate gold(I) alkynyl complexes with rod-like structure have been reported to show luminescence properties; namely, phosphinogold(I) alkynyl $[R'_3PAu(C\equiv C-R)]$, isocyanogold(I) alkynyl $[(R'-N\equiv C)Au(C\equiv C-R)]$ and dialkynylaurate(I) $[(R-C\equiv C)Au(C\equiv C-R)]^-$.

The gold(I) alkynyl system supported by an auxiliary phosphine ligand represents a major class of luminescent gold(I) complexes. There are two common literature methods to prepare the mononuclear phosphinogold(I) alkynyl compounds. The first one employs the reaction of a gold(I) alkynyl polymer, $[Au(C\equiv C-R)]_\infty$, with the phosphine ligand, while the second involves the treatment of the organic terminal alkyne with the corresponding gold(I) phosphine precursor, $[(R'_3P)AuCl]$ in the presence of a base (Scheme 17). The luminescence properties of such mononuclear complexes have been investigated by the research groups of Che [129–135], Mingos [136], Yam [136–138] and Puddephatt [139, 140]. Excitation of the solid sample and solution of these complexes gave rise to long-lived intense luminescence with vibronic structures. In principle, the complexes bearing aryl phosphine ligands with lower $\pi^*(PR'_3)$ orbital energies would display emission originating from $^3[\sigma(Au)\rightarrow\pi^*(PR'_3)]$ and metal-perturbed $^3[\pi\rightarrow\pi^*(C\equiv C-R)]$ excited states. On the other hand, only the triplet state of $[\pi\rightarrow\pi^*(C\equiv C-R)]$ character was suggested to be responsible for the emission of the complexes with alkyl phosphine groups [133, 134, 140]. It was suggested that introduction of the $[(R'_3P)Au]$ motif into the organic alkynyl backbone would enhance spin-orbit coupling from the heavy atom effect and improve the possibility to observe the alkynyl triplet emission at ambient temperature. Recently, Che and coworkers reported the luminescence properties of some mononuclear gold(I) complexes bearing electron-withdrawing alkynyl groups, $[Cy_3PAu(C\equiv C-R)](R=C_6H_4-NO_2-4$ **61**, $C_6H_4-CF_3-4$ **62**, C_6F_5 **63**, $C_6H_4-C_6H_5-4$ **64**) [132]. In contrast to the assignment of emission origin of **62–64** as a $^3[\pi\rightarrow\pi^*(C\equiv C-R)]$ excited state, the emission origin of **61** was suggested to be derived from intra-ligand charge transfer (ILCT) state. Such a difference in assignment was made on the basis of the observation of a structureless emission band and the solvent sensitivity of the emission energy of **61** in fluid solution. The extension of π-conjugation of the phenylacetylide moiety through the introduction of the NO_2 unit may

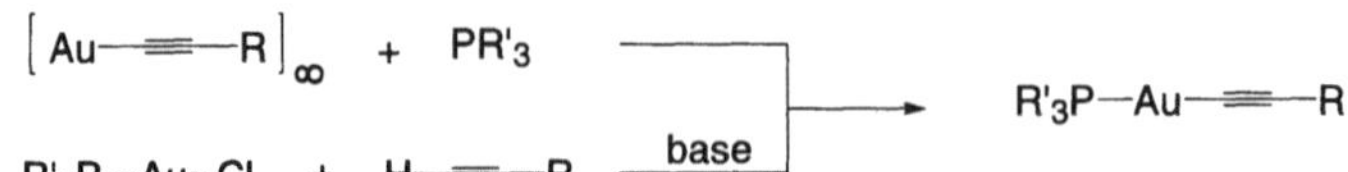

Scheme 17 Synthetic routes of phosphinogold(I) alkynyl complexes

account for such disparity. It is interesting to note that two polymorphs of **61** have been isolated and structurally characterized. The crystal packing revealed the differences of molecular orientation and dihedral angles between the 4-nitrophenyl groups in these two forms.

Dinuclear gold(I) alkynyl complexes [$R_3PAu(C\equiv C)Au(PR_3)$] [136] (Fig. 7) and **65–74** [133–136, 138] (Fig. 8) have also been synthesized by using a similar synthetic method with various bridging alkynyl ligands; all of which displayed intense luminescence. The emission bands of **65** and **66** were vibronically structured and were assigned to originate from metal-perturbed intra-ligand [$\pi \rightarrow \pi^*(C\equiv C-C_6H_4-C\equiv C)$] and [$\pi \rightarrow \pi^*(C\equiv C-C_{14}H_8)-C\equiv C)$] excited states, respectively. Lower emission energy was observed in **65** than that in its corresponding mononuclear analogue. Such a red shift in emission energy is due to the more extensive electron delocalization in the bridging alkynyl ligand. Interestingly, Che and coworkers reported the observation of dual emissions in **67–70** [135]. Identical excitation spectra were recorded for the high-energy and low-energy emission bands, suggesting that they are from the same absorbing state. Well-resolved vibronic fine structures were detected in the excitation spectra and were identified as the ground state phenyl ring deformation, symmetric phenyl ring stretch and $C\equiv C$ stretch. The low-energy emission was attributed to the $^3[\pi \rightarrow \pi^*]$ excited states of the bridging alkynyl, while the high-energy emission with microsecond lifetime was tentatively assigned to "delayed fluorescence" as a result of triplet-triplet annihilation. On the other hand, the dinuclear complexes **71–74** exhibited intense emissions, with sharp vibronic structures with progressional spacings corresponding to the $\nu(C\equiv C)$ stretch, assignable to the acetylenic $^3[\pi \rightarrow \pi^*]$ excited state localized on the $(C\equiv C)_n^{2-}$ chain [133, 134]. The "switching on" of the $^3[\pi \rightarrow \pi^*]$ emission of the $(C\equiv C)_n^{2-}$ moiety was facilitated by the ligation of [Cy_3PAu] through the spin-orbit coupling. Similar

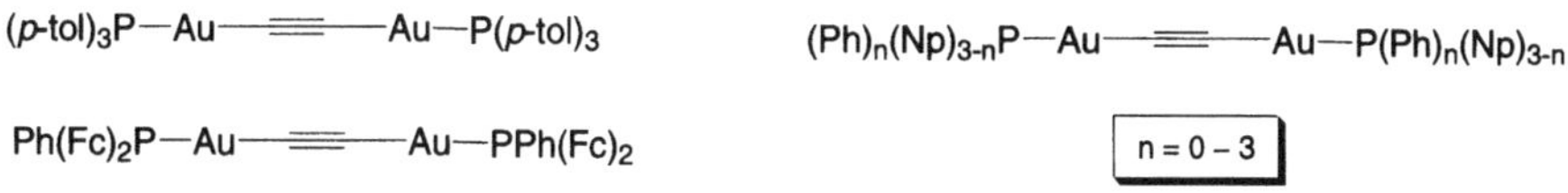

Fig. 7 Dinuclear ethynyl-bridged gold(I) phosphine complexes

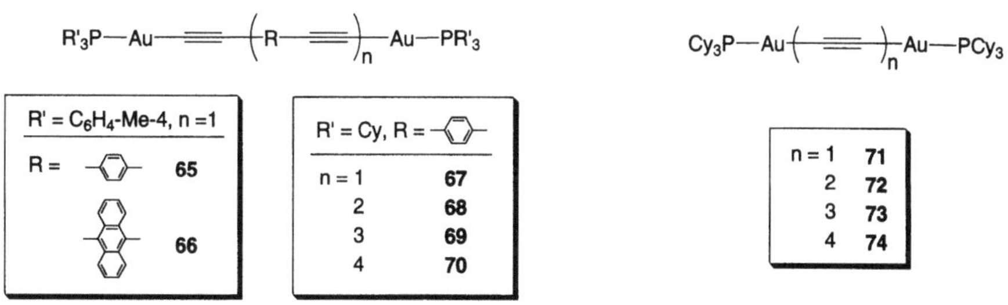

Fig. 8 Dinuclear oligoynyl-bridged gold(I) phosphine complexes of **65–74**

to the works of estimating the limiting absorption value for the lowest absorption of $R(C\equiv C)_{\infty}R$ by Hirsch [141] and Gladysz [71, 142], the limiting phosphorescence values for $R(C\equiv C)_{\infty}R$ and $R(C\equiv C-C_6H_4-C\equiv C)_{\infty}R$ have also been estimated by extrapolation of the infinite repeating units in the series of **67–70** and **71–74** from the plot of $\nu_{0\text{-}0}$ energy (the lowest triplet state energy) versus the reciprocal of chain length, $1/n$ (n = number of repeating alkynyl units) [133–135]. The energy gap for the singlet-triplet splitting of the $R(C\equiv C)_{\infty}R$ and $R(C\equiv C-C_6H_4-C\equiv C)_{\infty}R$ has been obtained from the values of limiting lowest triplet state energy.

The luminescence behaviors of linear gold(I) alkynyls with isocyanide ligand, $[(R'-N\equiv C)Au(C\equiv C-R)]$, have also been studied by Che [143] and Puddephatt [140]. The emission bands in the green were vibronically structured in both the solution state and 77-K glass. Similarly, assignment of a metal-modified intra-ligand $^3[\pi\rightarrow\pi^*(C\equiv C-R)]$ excited state was suggested for this emission. It is noteworthy that lower energy emission was observed in the solid state at room temperature, relative to that in the solution state and 77-K glass. Such a red shift in the emission energy was attributed to the presence of short Au – Au contacts as revealed from the crystal packing. Accordingly, the low-energy emission origin was assigned as derived from a metal-centered $^3[(d_\sigma^*)^1(p_\sigma)^1]$ excited state. Similar findings and assignments have also been reported for the related dinuclear gold(I) complex, $[\{Au(C\equiv CPh)\}_2(\mu-\text{dppe})]$ [129]. Dialkynylaurate(I) complexes, $[(R-C\equiv C)Au(C\equiv C-R)]^-$ belong to another class of linear gold(I) alkynyls that shows luminescence. Two alkynyl groups are attached to the gold(I) atom in a linear array to form an anionic molecular species. Actually, less exploration has been made for such complexes, with rare examples [130, 144–146]. The emission band observed in $[NBu_4][Au(C\equiv CPh)_2]$ was very similar to that of the mono-alkynyl analogue, $[(PPh_3)Au(C\equiv CPh)]$ and was ascribed to the metal-perturbed $^3[\pi\rightarrow\pi^*(C\equiv CPh)]$ state [130]. Not surprisingly, the emission origin was assigned as a metal-perturbed $^3[\pi\rightarrow\pi^*(C\equiv C-R)]$ emission.

6 Mixed-Metal

Since the rhenium(I) alkynyl complexes are air-, thermal- and photostable and are very robust in various environments due to the presence of the strong rhenium(I)–carbon bond, such classes of metal alkynyl complexes would serve as ideal candidates for the construction of heterometallic molecules by the concept of the "metalloligand" approach. In order to construct another metal–carbon bond in the target heterometallic molecules, a prerequisite for the rhenium(I) alkynyl precursor would be the availability of a terminal acetylenic group for further ligation. Two types of heterometallic alkynyl complexes, rhenium(I)-copper(I)/silver(I) and rhenium(I)-iron(II) systems, have been described and the complexes $[Re(CO)_3(N-N)(C\equiv C-C\equiv C-H)]$ and $[Re(CO)_3(N-N)(C\equiv C-C_6H_4-C\equiv C-H)]$ were employed as the rhenium(I) alkynyl precursors.

6.1 Rhenium(I)-Copper(I)/Silver(I)

With recent interest in both the polynuclear copper(I)/silver(I) and rhenium(I) alkynyl complexes, the combination of rhenium(I)-copper(I) and -silver(I) alkynyl systems would be expected to show interesting luminescence properties. With appropriate design and the judicious choice of ligands, preparation of heterometallic molecular rods **75–79** has been achieved by Yam and coworkers by employing a synthetic method that involves the use of a rhenium(I) alkynyl precursor as the metalloligand (Scheme 18) [44, 147, 148]. Excitation of **75–79** resulted in a strong orange luminescence in the solid state and in fluid solution and the emission energy was found to be dependent on the nature of the diimine and diphosphine ligands as well as the triangular metal core. With reference to previous luminescence studies of rhenium(I) alkynyl systems, this low-energy emission was assigned as derived from an excited state of a metal cluster modified ^{3}MLCT $[d\pi(Re)\rightarrow\pi^*(N-N)]$ origin. By the comparison of two similar copper(I) complexes **75** and **76**, with the same diphosphine ligand but different diimine ligand, the emission energy of **76** was found to be higher than that of **75**, which is in line with the ^{3}MLCT $[d\pi(Re)\rightarrow\pi^*(N-N)]$ assignment due to higher π^* orbital energy of ^{t}Bu$_2$bpy than bpy as a result of the electron-donating effect of the *tert*-butyl groups. For the complexes with the same diimine ligand, the occurrence of a slightly higher energy emission band of **77** relative to **78** was attributed to the more electron-deficient dppa bridging ligand, which would render the Cu_3 core more electron-withdrawing, resulting in a lower $d\pi$(Re)

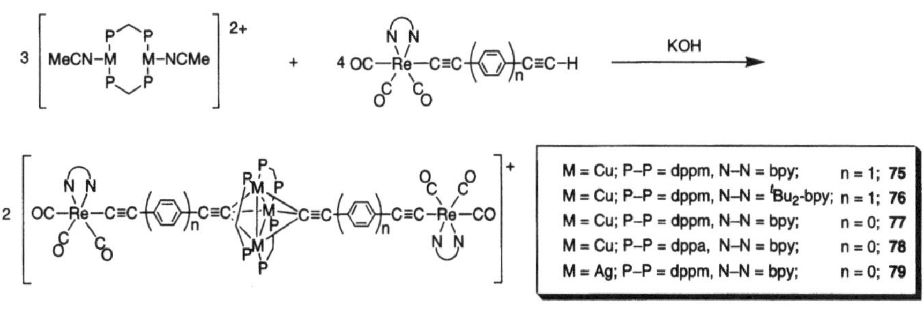

Scheme 18 Synthetic routes of heterometallic pentanuclear rhenium(I)-copper(I)/silver(I) complexes **75–79**

orbital energy and a higher ^{3}MLCT energy. On the other hand, higher emission energy was observed in **79** compared with that in **75**. Since copper(I) is more easily oxidized than silver(I), this implies that the Ag_3 core in **79** is more electron-withdrawing than the Cu_3 core in **75**. Similarly, the more electron-withdrawing Ag_3 core would lower the $d\pi$(Re) orbital energy and lead to a higher ^{3}MLCT energy.

6.2 Rhenium(I)-Iron(II)

In view of the rich redox and electrochemical behavior of iron(II) alkynyl system $[(C_5Me_5)(dppe)Fe(C\equiv C-R)]$ [25], a combination of the rhenium alkynyl with the iron system may lead to interesting perturbation of the luminescence properties of the rhenium(I) alkynyl system through the coordination of the iron(II) moiety and the capability of its oxidation to iron(III). A heterobimetallic iron(II)-rhenium(I) alkynyl complex, **80**, has been synthesized by the combination of the strongly emissive $[Re(diimine)(CO)_3(C\equiv C-R)]$ fragment and the redox active $[(C_5Me_5)(dppe)Fe(C\equiv C-R)]$ moiety, as reported by Lapinte, Yam and coworkers (Scheme 19) [149]. Complex **80**, unlike the rhenium(I) alkynyl precursor that showed intense luminescence, was found to be non-emissive. The lack of luminescence behavior in **80** was attributed to intramolecular reductive electron transfer and energy transfer quenching of the emissive ^{3}MLCT $[d\pi(Re)\rightarrow\pi^*(N-N)]$ excited state by the lower-lying MLCT and metal-centered LF states of the $[(C_5Me_5)(dppe)Fe(C\equiv C-R)]$ moiety. Com-

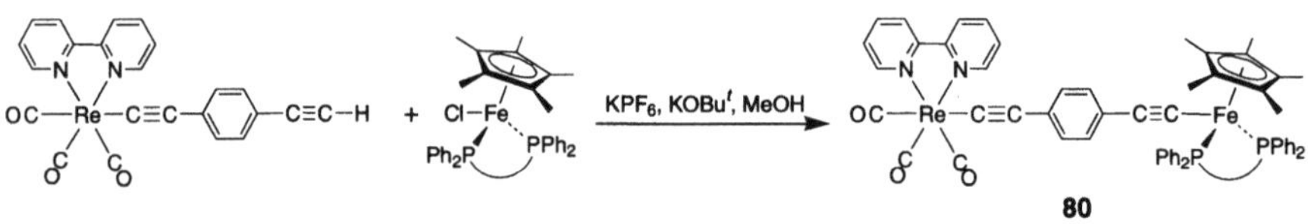

Scheme 19 Synthetic route of heterobimetallic iron(II)-rhenium(I) alkynyl complex **80**

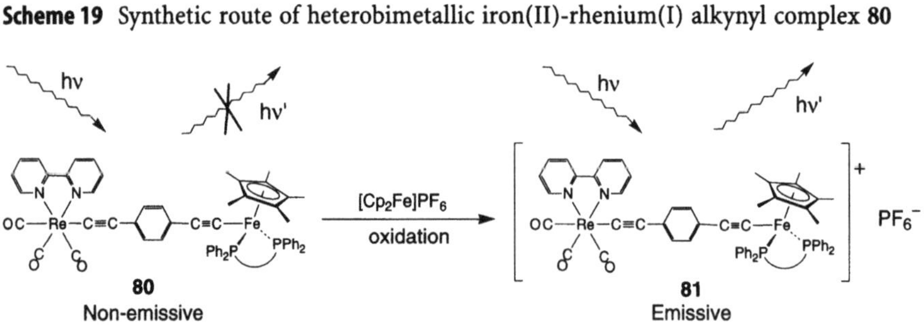

Scheme 20 Redox driven luminescence switching of heterobimetallic iron(II)-rhenium(I) alkynyl complex

plex **81**, the respective one-electron oxidized cationic species of **80**, was obtained and isolated by using the ferrocenium salt as an oxidizing agent (Scheme 20). It is interesting to note that **81** showed a recovery of the ^{3}MLCT $[d\pi(\mathrm{Re}) \rightarrow \pi^*(\mathrm{N-N})]$ emission, which was ascribed to the absence of reductive electron transfer quenching and energy transfer quenching processes resulting from the oxidation of iron(II) to the electron-deficient iron(III) center, which also led to the removal of the MLCT state of the $[(C_5Me_5)(\mathrm{dppe})\mathrm{Fe}(C \equiv C-R)]$ moiety and the raising of the energy of the LF state in the iron moiety.

6.3
Gold(I)-Rhenium(I)

Using 4-ethynylpyridine as a bridging ligand to tether two different transition-metal centers in its terminal alkynyl as well as the pyridyl group is another approach to prepare mixed-metal complexes. A series of mixed-metal gold(I)-rhenium(I) alkynyl complexes, $[PR_3Au(C \equiv C-C_5H_4N)Re(N-N)(CO)_3]^+$, have been synthesized by the reaction of the 4-ethynylpyridine gold(I) precursor complex with $[Re(N-N)(CO)_3(MeCN)]^+$ by Yam and coworkers (Scheme 21) [150]. These gold(I)-rhenium(I) alkynyl complexes were found to be emissive both in the solid state and in dichloromethane solution. An assignment of an alkynylgold-perturbed ^{3}MLCT $[d\pi(\mathrm{Re}) \rightarrow \pi^*(\mathrm{N-N})]$ origin has been made for such emission.

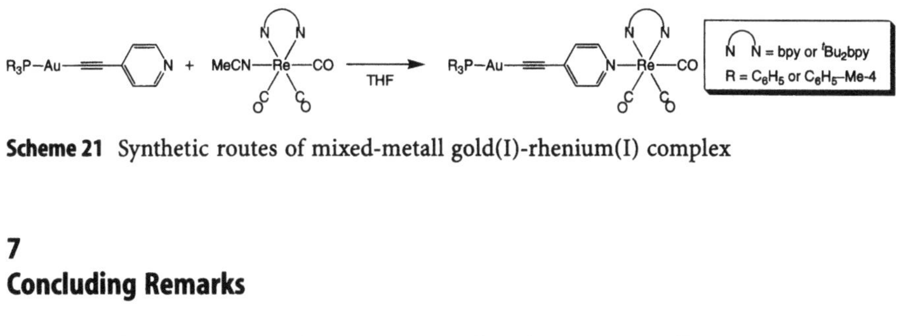

Scheme 21 Synthetic routes of mixed-metall gold(I)-rhenium(I) complex

7
Concluding Remarks

In view of the presence of *sp*-unsaturated carbon atoms, alkynyl groups with different bonding fashions may serve as versatile building blocks in the construction of alkynyl transition-metal complexes. In this review, a variety of luminescent alkynyl systems of different transition-metal centers, such as rhenium(I), ruthenium(II), platinum(II), copper(I), silver(I), and gold(I) as well as their mixed-metal complexes with one-dimensional molecular rod structure are presented. The synthesis and luminescence properties have also been described. Upon introduction of an alkynyl group, the luminescence behavior would be perturbed via the modification of the lowest-energy excited states and as a consequence the emission color could be fine-tuned and the luminescence origins varied. Extension of the work to luminescence biolabeling using luminescent metal alkynyl complexes has also been accomplished with specifically functionalized alkynyl groups. Through rational design, a redox-driven luminescence switch has been constructed in a heterometallic alkynyl complex by the incorporation of a redox active metal center that is capable of switching via chemically induced oxidation. Such luminescent metal alkynyl complexes may provide interesting new opportunities for the construction of molecular devices with tunable luminescence behavior.

Acknowledgements V.W.-W.Y. acknowledges support from the University Development Fund of the University of Hong Kong, the University of Hong Kong Foundation for Educational Development and Research Limited, and a CERG grant from the Research Grants Council of the Hong Kong Special Administrative Region, China (Project No. HKU 7123/00P).

References

1. Horváth O, Stevenson KL (eds) (1993) Charge transfer photochemistry of coordination compounds. VCH, New York
2. Balzani V, Scandola F (eds) (1991) Supramolecular photochemistry. Ellis Horwood, Chichester, UK
3. Kalyanasundaram K (1992) Photochemistry of polypyridine and porphyrin complexes. Academic press, London
4. Solomon EI, Lever ABP (eds) (1999) Inorganic electronic structure and spectroscopy, Vol. II. Wiley, New York
5. Hagfeldt A, Grätzel M (1995) Chem Rev 95:49
6. Bignozzi CA, Argazzi R, Kleverlaan CJ (2000) Chem Soc Rev 29:87
7. Xie PH, Hou YJ, Zhang BW, Cao Y, Wu F, Tian WJ, Shen JC (1999) J Chem Soc Dalton Trans 4217
8. Klein C, Nazeeruddin MK, Censo DD, Liska P, Grätzel M (2004) Inorg Chem 43:4216
9. Baldo MA, O'Brien DF, Thompson ME, Forrest SR (1999) Phys Rev B 60:14422
10. Baldo MA, Thompson ME, Forrest SR (2000) Nature 403:750
11. Welter S, Brunner K, Hofstraat JW, De Cola L (2003) Nature 421:54
12. Xin H, Li FY, Shi M, Bian ZQ, Huang CH (2003) J Am Chem Soc 125:7166
13. Chan SC, Chan MCW, Che CM, Wang Y, Cheung KK, Zhu N (2001) Chem Eur J 7:4180
14. Lu W, Mi BX, Chan MCW, Hui Z, Che CM, Zhu N, Lee ST (2004) J Am Chem Soc 126:4958
15. Tour JM (2000) Acc Chem Res 33:791
16. Kim J, Swagar TM (2001) Nature 411:1030
17. Martin RE, Diederich F (1999) Angew Chem Int Ed 38:1350
18. Marder SR, Sohn JE, Stucky GD (eds) (1991) Materials for nonlinear optics: chemical perspectives. ACS Symposium Series vol 455. American Chemical Society, Washington, DC
19. Bruce DW, O'Hare D (eds) (1992) Inorganic materials. Wiley, Chichester, U.K.
20. Long NJ (1995) Angew Chem Int Ed Engl 34:21
21. Humphrey MG, Powell CE (2004) Coord Chem Rev 248:725
22. Manna J, John KD, Hopkins MD (1995) Adv Organomet Chem 38:79

23. Long NJ, Williams CK (2003) Angew Chem Int Ed 42:2586
24. Beljonne D, Wittman HF, Köhler A, Graham S, Younus M, Lewis J, Raithby PR, Khan MS, Friend RH Brédas JL (1996) J Chem Phys 105:3868
25. Paul F, Lapinte C (1998) Coord Chem Rev 178–180:431
26. Houlding VH, Miskowski VM (1991) Coord Chem Rev 111:145
27. McMillin DR, Moore JJ (2002) Coord Chem Rev 229:113
28. Mihan S, Sunkel K, Beck W (1999) Chem Eur J 5:745
29. Lang H, Kohler K, Blau S (1995) Coord Chem Rev 143:113
30. Nast R (1982) Coord Chem Rev 47:89
31. Yam VWW, Yu KL, Cheung KK (1999) J Chem Soc Dalton Trans 2913
32. Abu-Salah OM, Al-Ohaly ARA (1998) J Chem Soc Dalton Trans 2297
33. Szafert S, Gladysz JA (2003) Chem Rev 103:4175
34. Wrighton MS, Morse DL (1974) J Am Chem Soc 96:998
35. Fredericks SM, Luong JC, Wrighton MS (1979) J Am Chem Soc 101:7415
36. Caspar JV, Meyer TJ (1983) J Phys Chem 87:952
37. Reitz A, Dressick WJ, Demas JN (1986) J Am Chem Soc 108:5344
38. Tapolsky G, Duesing R, Meyer TJ (1990) Inorg Chem 29:2285
39. Lin R, Fu Y, Brock CP, Guarr TF (1992) Inorg Chem 31:4346
40. Koike K, Okoshi N, Hori H, Takeuchi K, Ishitani O, Tsubaki H, Clark IP, George MW, Johnson FPA, Turner JJ (2002) J Am Chem Soc 124:11448
41. Yam VWW, Lau VCY, Cheung KK (1995) Organometallics 14:2749
42. Yam VWW, Chong SHF, Cheung KK (1998) Chem Commun 2121
43. Yam VWW, Lo KKW, Wong KMC (1998) J Organomet Chem 578:3
44. Yam VWW (2001) Chem Commun 789
45. Yam VWW, Chong SHF, Ko CC, Cheung KK (2000) Organometallics 19:5092
46. Yam VWW, Wong KMC, Chong SHF, Lau VCY, Lam SCF, Zhang L, Cheung KK (2003) J Organomet Chem 670:205
47. Yam VWW, Lau VCY, Cheung KK (1996) Organometallics 15:1740
48. Chong SHF (2000) PhD Thesis, The University of Hong Kong
49. Yam VWW, Chu BWK, Cheung KK (1998) Chem Commun 2261
50. Yam VWW, Chu BWK, Ko CC, Cheung KK (2001) J Chem Soc Dalton Trans 1911
51. Yam VWW, Ko CC, Chu BWK, Zhu N (2003) J Chem Soc Dalton Trans 3914
52. Wong CY, Chan MCW, Zhu N, Che CM (2004) Organometallics 23:2263
53. Sacksteder L, Baralt E, DeGraff BA, Lukehart CM, Demas JN (1991) Inorg Chem 30:2468
54. Choi CL, Cheng YF, Yip C, Phillips DL, Yam VWW (2000) Organometallics 19:3192
55. Lewis J, Khan MS, Kakkar AK, Johnson BFG, Marder TB, Fyfe HB, Wittmann F, Friend RH, Dray AE (1992) J Organomet Chem 425:165
56. Laine RM (ed) (1992) Fyfe HB, Mlekuz M, Stringer G, Taylor NJ, Marder TB Inorganic and organometallic polymers with special properties. NATO ASI Series. Kluwer Acad Publ, Dordrecht, The Netherlands, p331
57. Wittmann HF, Friend RH, Khan MS, Lewis J (1994) J Chem Phys 101:2694
58. Chawdhury N, Köhler A, Friend RH, Younus M, Long NJ, Raithby PR, Lewis J (1998) Macromolecules 31:722
59. Lewis J, Long NJ, Raithby PR, Shields GP, Wong WY, Younus M (1997) J Chem Soc Dalton Trans 4283
60. Wong WY, Wong WK, Raithby PR (1998) J Chem Soc Dalton Trans 2761
61. Lewis J, Raithby PR, Wong WY (1998) J Organomet Chem 556:219
62. Younus M, Köhler A, Cron S, Chawdhury N, Al-Mandhary MRA, Khan MS, Lewis J, Long NJ, Friend RH, Raithby PR (1998) Angew Chem Int Ed 37:3036

63. Wilson JS, Köhler A, Friend RH, Al-Suti MK, Al-Mandhary MRA, Khan MS, Raithby PR (2000) J Chem Phys113:7627
64. Köhler A, Wilson JS, Friend RH, Al-Suti MK, Khan MS, Gerhard A, Bässler H (2002) J Chem Phys 116:9457
65. Furlani A, Licoccia S, Russo MV, Villa AC, Guastini C (1982) J Chem Soc Dalton Trans 2449
66. Furlani A, Licoccia S, Russo MV, Billa AC, Guastini C (1984) J Chem Soc Dalton Trans 2197
67. Casalboni M, Sarcinelli F, Pizzoferrato R, D'Amato R, Furlani A, Russo MV (2000) Chem Phys Lett 319:107
68. D'Amato R, Furlani A, Colapietro M, Portalone G, Casalboni M, Falconieri M, Russo MV (2001) J Orgomet Chem 627:13
69. Liu Y, Jiang S, Glusac K, Powell DH, Anderson DF, Schanze KS (2002) J Am Chem Soc 124:12412
70. Stahl J, Bohling JC, Bauer EB, Peters TB, Mohr W, Martín-Alvarez JM, Hampel F, Gladysz JA (2002) Angew Chem Int Ed 41:1871
71. Mohr W, Stahl J, Hampel F, Gladysz JA (2003) Chem Eur J 9:3324
72. Pringle PG, Shaw BL (1982) J Chem Soc Dalton Trans 581
73. Zipp AP (1988) Coord Chem Rev 84:47
74. Smith DC, Gray HB (1990) Coord Chem Rev 100:169
75. Roundhill DM, Gray HB, Che CM (1989) Acc Chem Res 22:55
76. Yersin H, Humbs W, Strasser J (1997) Coord Chem Rev 159:325
77. Balch AL (1976) J Am Chem Soc 98:8049
78. Lewis NS, Mann KR, Gordon JG, Gray HB (1976) J Am Chem Soc 98:7461
79. Yam VWW, Yu KL, Wong KMC, Cheung KK (2000) Organometallics 20:721
80. Wong KMC, Hui CK, Yu KL, Yam VWW (2002) Coord Chem Rev 229:123
81. Yam VWW, Hui CK, Wong KMC, Zhu N, Cheung KK (2002) Organometallics 21:4326
82. Hui CK, Chu BWK, Zhu N, Yam VWW (2003) Inorg Chem 41:6178
83. Aldridge TK, Stacy EM, McMilin DR (1994) Inorg Chem 33:722
84. Bailey JA, Hill MG, Marsh RE, Miskowski VM, Schaefer WP, Gray HB (1995) Inorg Chem 34:4591
85. Büchner R, Cunningham CT, Field JS, Haines R, McMillin DR, Summerton GC (1999) J Chem Soc Dalton Trans 711
86. Büchner R, Field JS, Haines R, Cunningham CT, McMillin DR (1997) Inorg Chem 36:3952
87. Yip HK, Cheng LK, Cheung KK, Che CM (1993) J Chem Soc Dalton Trans 2933
88. Bailey JA, Miskowski VM, Gray HB (1993) Inorg Chem 32:369
89. Hill MG, Bailey JA, Miskowski VM, Gray HB (1996) Inorg Chem 35:4585
90. Arena G, Calogero G, Campagna S, Scolaro LM, Ricevuto V, Romeo R (1998) Inorg Chem 37:2763
91. Lai SW, Chan MCW, Cheung KK, Che CM (1999) Inorg Chem 38:4262
92. Yutaka T, Mori I, Kurihara M, Mizutani J, Tamai N, Kawai T, Irie M, Ishihara H (2002) Inorg Chem 41:7143
93. Yam VWW, Tang RPL, Wong KMC, Cheung KK (2001) Organometallics 20:4476
94. Yam VWW, Wong KMC, Zhu N (2003) Angew Chem Int Ed 42:1400
95. Yam VWW, Wong KMC, Zhu N (2002) J Am Chem Soc 12:6506
96. Wong KMC, Tang WS, Chu BWK, Zhu N, Yam VWW (2004) Organometallics 23:3459
97. Tzeng BC, Fu WF, Che CM, Chao HY, Cheung KK, Peng SM, (1999) J Chem Soc Dalton Trans 1017
98. Yam VWW, Tang RPL, Wong KMC, Ko CC, Cheung KK (2001) Inorg Chem 40:571

99. Osborn RS, Rogers D (1974) J Chem Soc Dalton Trans 1002
100. Herber RH, Croft M, Coyer MJ, Bilash B, Sahiner A (1994) Inorg Chem 33:2422
101. Yam VWW, Hui CK, Yu SY, Zhu N (2004) Inorg Chem 43:812
102. Lu W, Mi BX, Chan MCW, Hui Z, Zhu N, Lee ST, Che CM (2002) Chem Commun 206
103. Gamasa MP, Gimeno J, Lastra E, Aguirre A, Garcia-Granda (1989) J Organomet Chem 378:C11
104. Diéz J, Gamasa MP, Gimeno J, Aguirre A, Garcia-Granda (1991) Organometallics 10:380
105. Yam VWW, Lee WK, Lai TF (1993) Organometallics 12:2383
106. Yam VWW, Lee WK, Yeung PKY, Phillips D (1994) J Phys Chem 98:7545
107. Yam VWW, Fung WKM, Wong MT (1997) Organometallics 16:1772
108. Yam VWW, Fung WKM, Cheung KK (1998) Organometallics 17:3293
109. Yam VWW, Fung WKM, Cheung KK (1999) J Cluster Sci 10:37
110. Yam VWW (1997) J Photochem Photobiol A Chem 106:75
111. Yam VWW, Lo KKW, Fung WKM, Wang CR (1998) Coord Chem Rev 171:17
112. Lo WY, Lam CH, Yam VWW, Zhu N, Cheung KK, Fathallah S, Messaoudi S, Guennic BL, Kahlal S, Halet JF (2004) J Am Chem Soc 126:7300
113. Yam VWW, Fung WKM, Cheung KK (1997) Chem Commun 963
114. Yam VWW, Fung WKM, Cheung KK (1997) Organometallics 16:2032
115. Moore CE (1971) Natl Stand Ref Data Ser, US Natl Bur Stand, NSRDS-NBS 35:51&116
116. King C, Wang JC, Khan Md NI, Fackler JP (1989) Inorg Chem 28:2145
117. Che CM, Kwong HL, Yam VWW, Cho KC (1989) J Chem Soc Dalton Trans 885
118. Li D, Che CM, Kwong HL, Yam VWW (1992) J Chem Soc Dalton Trans 3325
119. Yam VWW, Lee WK (1993) J Chem Soc Dalton Trans 2097
120. Leung KH, Phillips DL, Tse MC, Che CM, Miskowski VM (1999) J Am Chem Soc 121:4799
121. Vogler A, Kunkely H (1988) Chem Phys Lett 150:135
122. Forward JM, Bohmann D, Fackler JP, Staples RJ (1995) Inorg Chem 34:6330
123. Mansour MA, Connick WB, Lachicotte RJ, Gysling HJ, Eisenberg R (1998) J Am Chem Soc 120:1329
124. Bickery JC, Olmstead MM, Fung EY, Balch AL (1997) Angew Chem Int Ed 36:1179
125. Yam VWW, Li CK, Chan CL (1998) Angew Chem 37:2857
126. Lee YA, McGarrah JE, Lachicotte RJ, Eisenberg R (2002) J Am Chem Soc 124:10662
127. White-Morris RL, Olmstead MM, Balch AL (2003) J Am Chem Soc 125:1033
128. White-Morris RL, Olmstead MM, Jiang F, Tinti DS, Balch AL (2002) J Am Chem Soc 124:2327
129. Li D, Hong X, Che CM, Lo WC, Peng SM (1993) J Chem Soc Dalton Trans 2929
130. Che CM, Yip HK, Lo WC, Peng SM (1994) Polyhedron 13:887
131. Shieh SJ, Hong X, Peng SM, Che CM (1994) J Chem Soc Dalton Trans 3067
132. Lu W, Zhu N, Che CM (2003) J Am Chem Soc 125:16081
133. Che CM, Choa HY, Miskowski VM, Li Y, Cheung KK (2001) J Am Chem Soc 123:4985
134. Lu W, Xiang HF, Zhu N, Che CM (2002) Organometallics 21:2343
135. Chao HY, Lu W, Chan MCW, Che CM, Cheung KK, Zhu N (2002) J Am Chem Soc 124:14696
136. Müller TE, Choi SWK, Mingos DMP, Murphy D, Williams DJ, Yam VWW (1994) J Organomet Chem 484:209
137. Yam VWW, Choi SWK, Cheung KK (1996) Organometallics 15:1734
138. Yam VWW, Choi SWK, Cheung KK (1996) J Chem Soc Dalton Trans 4227
139. Hunks WJ, MacDonald MA, Jennings MC, Puddephatt RJ (2000) Organometallics 19:5063

140. Irwin MJ, Vittal JJ, Puddephatt RJ (1997) Organometallics 16:3541
141. Schermann G, Grösser T, Hampel F, Hirsch A (1997) Chem Eur J 3:1105
142. Dembinski R, Bartik T, Jaeger M, Gladysz JA (2000) 122:810
143. Xiao H, Cheung KK, Che CM (1996) J Chem Soc Dalton Trans 3699
144. Vicent J, Chicote MT, Abrisqueta MD, Falcón-Alvarez AF (2002) J Organomet Chem 663:40
145. Ferrer M, Rodríguez, Rossell O, Pina F, Lima JC, Mardia MF, Solans X (2003) J Organomet Chem 678:82
146. Yamanoto Y, Shiotsuka M, Okuno S, Onaka S (2004) Chem Lett 33:210
147. Yam VWW, Fung WKM, Wong KMC, Lau VCY, Cheung KK (1998) Chem Commun 777
148. Yam VWW, Lo WY, Lam CH, Fung WKM, Wong KMC, Lau VCY, Zhu N (2003) Coord Chem Rev 245:39
149. Wong KMC, Lam SCF, Ko CC, Zhu N, Yam VWW, Roué S, Lapinte C, Fathallah S, Costuas K, Kahlal S, Halet JF (2003) Inorg Chem 42:7086
150. Cheung KL, Yip SK, Yam VWW (2004) J Organomet Chem 689:4451

Top Curr Chem (2005) 257: 33–62
DOI 10.1007/b136066

Published online: 6 July 2005

Molecular Wires

Dustin K. James[1] (✉) · James M. Tour[2]

[1]Chemistry Department, Rice University, MS-60, 6100 Main St., Houston, Texas 77005, USA
dustin@rice.edu

[2]Chemistry Dept. and Center for Nanoscale Science and Technology, Rice University, MS-222, 6100 Main St., Houston, Texas 77005, USA
tour@rice.edu

1 **Introduction** 34

2 **Molecular Electronics** 34
2.1 Optoelectronics 36

3 **Molecular Wires** 36
3.1 Organic Molecular Wires 37
3.1.1 Oligo(2,5-thiophene ethynylene)s 37
3.1.2 Oligo(1,4-phenylene ethynylene)s 38
3.1.3 Oligo(1,4-phenylene vinylene)s 43
3.1.4 Aromatic Ladder Oligomers 46
3.1.5 Oligophenylenes and Polyphenylenes 47
3.1.6 Acetylene Oligomers 49
3.1.7 Carbon Nanotubes 50
3.2 Organometallic Molecular Wires 51

4 **Measurement of Conduction in Molecular Wires** 55

5 **Conclusion** 58

References 58

Abstract Molecular wires are compounds that are proposed to be used in molecular electronic and optoelectronic devices to replace the metal and silicon-based wires in semiconductor devices. We review the field, including organic molecular wires such as oligo(2,5-thiophene ethynylene)s, oligo(1,4-phenylene ethynylene)s, oligo(1,4-phenylene vinylene)s, aromatic ladder oligomers, oligophenylenes, polyphenylenes, acetylene oligomers, carbon nanotubes, and organometallic molecular wires. We briefly review the measurement of conduction in molecular wires and conclude that fully conjugated organic aromatic molecular wires are the best candidates for introduction into new electronic devices as replacements for the Al or Cu wiring presently used in logic and memory devices.

Keywords Molecular wires · Molecular electronics · Optoelectronic · Conductance · Self-assembly · Calculation

Abbreviations

cAFM	conducting atomic force microscopy
CNTFETs	carbon nanotube field effect transistors
CV	cyclic voltammetry
DMSO	dimethylsulfoxide
NDR	negative differential resistance
OLED	organic light emitting diodes
OTEs	oligo(2-5-thiophene ethynylene)s
OPEs	oligo(1,4-phenylene ethynylene)s
OPVs	oligo(1,4-phenylene vinylene)s
SAM	self-assembled monolayer
STM	scanning tunneling microscopy
SWE	single wavelength ellipsometry
SWNT	single-walled nanotube
Terpy	terpyridine
XPS	X-ray photoelectron spectroscopy

1 Introduction

Molecular wires have as their basis real-world examples. Metallic wiring 1 cm in diameter facilitates the flow of electrons that power our household lighting, refrigerators and other appliances, entertainment electronics such as televisions and stereos, and relatively new electronic devices such as computers. Within those electronic devices, wires 1 mm wide on printed circuit boards connect resistors, rheostats, and logic chips. Inside those logic chips, wires tenths of a μm wide connect solid-state transistors, carved out of silicon, and allow them to act in concert with thousands of similar transistors to carry out computations. This last size reduction nearly reaches what is thought to be the limit of present semiconductor manufacturing technology. To enable further miniaturization, recent research has produced molecular-scale wires, ranging in length from 1 to 100 nm and width from ~ 0.3 nm on up. In this chapter we will give an overview of the synthesis and characterization of a number of classes of molecular wires.

2 Molecular Electronics

The rapidly developing field of molecular electronics is one of the driving forces behind the interest in molecular wires [1–8]. In our recent reviews we covered both the synthetic aspects of molecular wires as well as the large body of work concerning the theoretical aspects of conduction by molecular wires [1, 9]; Robertson reviewed the field in 2003 [10]. The limitations of the present "top-down" method of producing semiconductor-based devices have been the subject of debate and conjecture since Gordon Moore's prediction that the number of components per integrated circuit would double every 18 months [11]. It was thought that the inherent limitations of the present technology would lead to a dead-end in the next few years. For instance, silicon's band structure disappears when silicon layers are just a few atoms thick. Lithographic techniques that are used to produce the circuitry on the silicon wafers are limited by the wavelengths at which they work. However, leaders in the semiconductor manufacturing world are still making advances that appear to be pushing "Moore's Law" beyond its prior perceived limits. Intel has declared that Moore's Law is here to stay for the next 15–20 years [12]. In the commercial technology of 2004, the copper wires in Intel's Pentium® 4 logic chip being made in their newest 300 mm wafer fabrication facility in Ireland are 90 nm wide [13]. Strained silicon [14] is but one of several approaches taken by the industry to modify their present silicon-based processes to meet the demands of the development roadmap.

For comparison's sake, a typical molecular wire synthesized in our lab is calculated to be 0.3 nm wide and 2.5 nm in length, Fig. 1 [4]. It would take 300 of these molecules, side-by-side, to span the 90 nm metal line in the most advanced logic chip being made today. The small size of these molecules is emphasized when one considers that 500 g (about one mole) of this wire would contain 6×10^{23} molecules, or more molecules than the number of transistors ever made in the history of the world. This amount of mate-

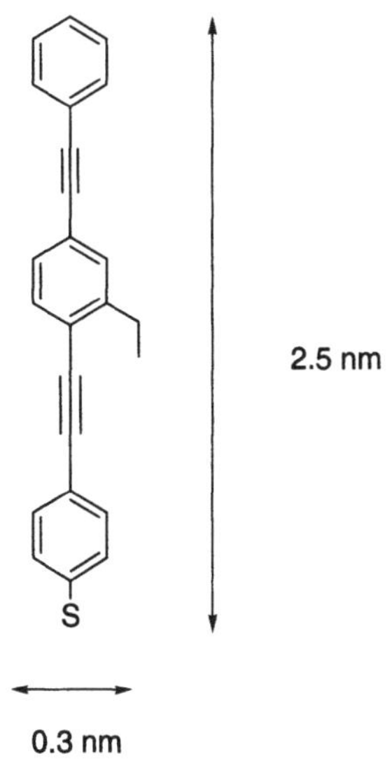

Fig. 1 The dimensions of a typical molecular wire are calculated to be 0.3 nm in width and 2.5 nm in length using molecular mechanics (Spartan) to determine the energy minimized structure [4]

rial could be produced using relatively small 22 L laboratory reaction flasks. Changing the physical characteristics of this wire is as easy as changing the raw materials used to make it. The small size, the potential of synthesizing huge numbers in small reactors, and the ease of modification of the physical characteristics of the molecules are good reasons for pursuing molecular wire research. As an example of how far the technology has come, molecular electronics is discussed in the "emerging research devices" section of the most recent International Technology Roadmap for Semiconductors [15, 16] and new molecular wires are a large part of the emerging technology.

2.1
Optoelectronics

Due to their chemical structure, some highly conjugated molecular wires have applications in optoelectronics [17–19]. Poly(phenylene vinylene)s are being used as components in organic light-emitting diodes (OLEDs) in displays such as used in cell phones and other electronic devices. Various other polymeric materials and small molecules are also being used or in development. These materials are applied in very thin layers about 100 nm thick, with the organic small molecules forming crystalline phases. This is quite different from the molecular electronics field, where it is envisioned that single molecules will eventually be used in circuits. However, much of the literature we will review addresses the optoelectronics applications of the molecular wires, and so we have included leading references for that area of research.

3
Molecular Wires

In this chapter, when we say "molecular wires" we mean discrete molecules, not crystals or films. We accept Cotton's statement that linear chains of metal atoms that exist only in the solid-state via stacking of flat molecules, or through formation of μ-bridged chains of octahedral molecules (such as $NbCl_4$) should not be called molecular wires [20]. The extremely interesting inorganic crystalline nanowires being developed by Lieber and others [21–24] may eventually be used as wiring in molecular electronics-based circuitry, but the fact that they are comprised of crystalline phases and not discrete molecules precludes their inclusion in this review.

In our survey of the literature, we find two general types of molecular wires. The largest portion of the literature, including most of our work, covers organic molecular wires. A smaller portion of the literature covers organometallic molecular wires. We include what some may call inorganic molecular wires in the organometallic class due to ease of classification and the small number of inorganic molecular wires in the literature.

Molecular wires are meant to conduct electricity between two points of a circuit. However, a majority of the molecular wires intended for molecular electronics have never been tested in an actual circuit. One reason for this is that it is difficult to do so because of their small size. Another reason is that there is not one generally accepted testbed that is readily available to all researchers. Rather there are several different testbeds in the literature [25–28], which we recently reviewed [29], and the results from those molecular wires that have been tested are in many cases not comparable. These devices are difficult to make, yields are low, and obtaining reproducible results requires care and patience. Drawing conclusions about the activity of classes of compounds or building structure activity relationships among several classes using the data generated can be a difficult exercise. However, since it has been shown that aromatic thiolates are much higher conducting that alkane thiolates [30] when bonded to Au surfaces, much attention has focused on conjugated aromatic molecular wires (vida infra).

3.1 Organic Molecular Wires

3.1.1 Oligo(2,5-thiophene ethynylene)s

Our group has focused on the synthesis of organic molecular wires. One class of compounds synthesized are the oligo(2,5-thiophene ethynylene)s (OTEs), several examples of which are compounds **1–5** as shown in Fig. 2 [31–34]. This class of rigid-rod oligomeric molecular wires was made through an iterative divergent-convergent synthesis method that allowed the quick assembly of the products, doubling their length at each step. The longest molecular wire synthesized was 12.8 nm in length. Note that these wires have thioester groups at one or both ends. When deprotected in-situ, the thiol groups enable the molecular wires to adhere to Au (or other metal) surfaces [30], therefore serving as "alligator clips". When large numbers of molecular wires bond to Au in a regular, packed array, through this self-assembly process, the group of molecules is called a self-assembled monolayer, or SAM. The bonding of the S atom to Au enables the flow of electricity from the Au metal Fermi levels through the S molecular orbitals to the molecular orbitals formed by the conjugated portion of the molecule. The ethynyl units in between the aromatic molecules are used in order to maintain maximum overlap of the orbitals, and to keep the molecules in a rod-like shape. The various side chains appended to the thiophene cores were intended to increase the organic solvent solubility of the wires. Unfunctionalized rigid-rod oligomers of this length suffer from severe solubility problems.

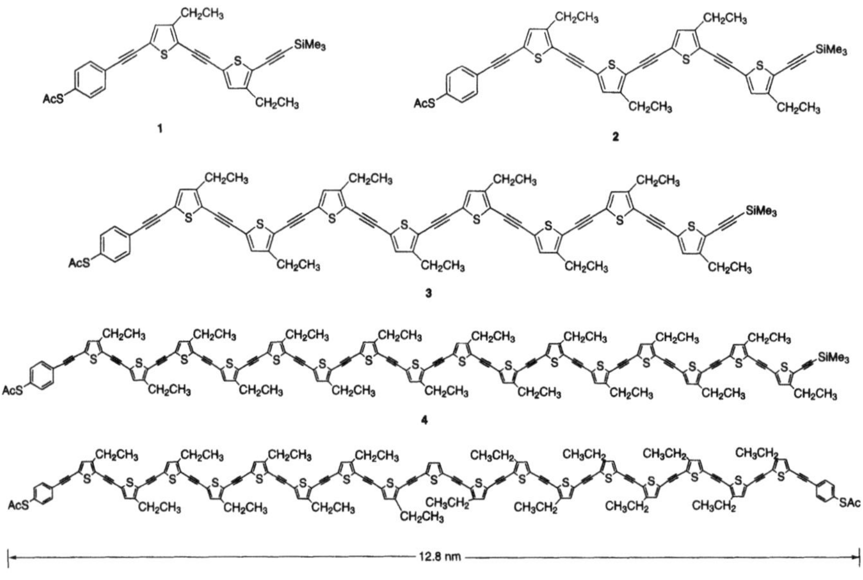

Fig. 2 Oligo(2,5-thiophene ethynylene) molecular wires **1–5** synthesized by us [31]. Note that **1–4** have thioester termini on one end only, while molecular wire **5** has thioester termini on both ends. The length of **5** is 12.8 nm for its energy-minimized conformation

3.1.2 Oligo(1,4-phenylene ethynylene)s

A second class of molecules that has been studied extensively in our lab [35] and by others [36, 37] are the oligo(1,4-phenylene ethynylene)s (OPEs). The molecule shown in Fig. 1 is of this class, as are the molecules shown in Fig. 3. As with the OTEs, the OPEs can be rapidly synthesized using transition-metal-catalyzed coupling reactions. In this case the compounds were synthesized in both solution phase and on a polymer-based solid resin. Compounds **6–9** of Fig. 3 are intermediates that were produced by cleaving the products from the resin using iodomethane and coupling the resulting aryl iodides to the alligator clip acetyl (4-ethynylthiophenol) using typical transition-metal-catalyzed coupling reactions [35]. Removal of the trimethylsilyl protecting groups from the alkynes produced compounds **10–12** (from **6**, **7**, and **9**, respectively), and coupling of the terminal alkyne in each case to acetyl (4-iodothiophenol) produced compounds **13** and **14** (from **10** and **12**, respectively). As in the OTEs, the C_{12} side chains were used to impart organic solubility to the products. The use of longer side chains such as C_{16} can result in side chain interdigitation, leading to insolubility problems rather than increasing solubility.

In a series of syntheses of molecules in OPE subclasses, we have made products with more than 2 terminals and derivatives that are meant to test the

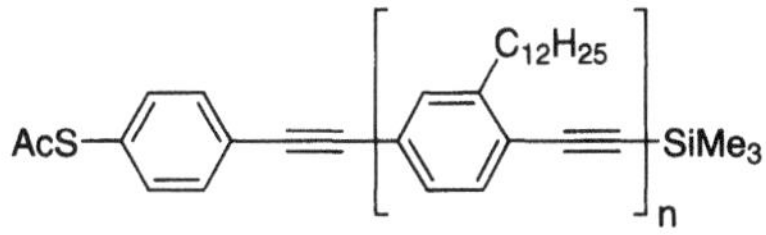

6, n = 1
7, n = 4
8, n = 8
9, n = 16

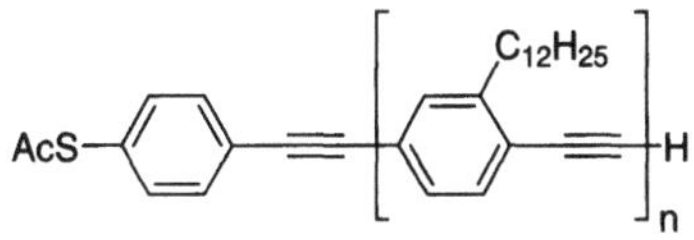

10, n = 1
11, n = 4
12, n = 16

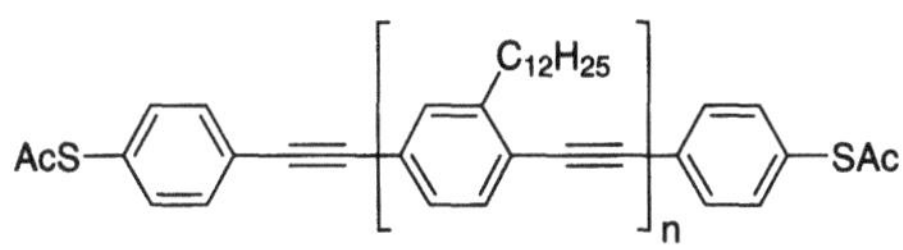

13, n = 1
14, n = 16

Fig. 3 A series of oligo(phenylene ethynylene)s (OPEs) were synthesized using a polymer support to increase the yields and facilitate purification [35]

necessity for a completely conjugated system [38]. The unfortunate circumstance is that presently no reliable system exists for testing 3- and 4-terminal molecular wires. However, the conductance of alkanethiolates (containing no aromatic conjugation whatsoever) on Au surfaces has been determined and shown to be less than that of conjugated systems (vida infra).

To further explore the organic functionality necessary for molecular wires to carry current, we synthesized a group of 2-terminal molecular wires **15–20**, shown in Fig. 4, that contain interior methylene or ethylene group barriers to electrical conduction, and that could be tested using presently known testbeds [38]. Each of these was synthesized using relatively straightforward chemistry, a fact that illustrates our earlier claim that it is easy to explore molecular wire space by changing just one or two aspects of the synthesis. We also synthesized a series of OPEs with different alligator clips to see what effect that variation would have on the conductance of the molecular wire [39] and we have developed combinatorial chemistry routes that are capable of the synthesis of 10s to 100s of new molecular wires at one time [40].

When members in the OPE family were functionalized with groups other than aliphatic ones, see Fig. 5, we began to see switching behavior [26,41,42], although there is some question about whether the switching behavior is due to conformational changes [41] or to the effects of changing functionality contained in the molecule of interest [26]. In the nanopore [26], mo-

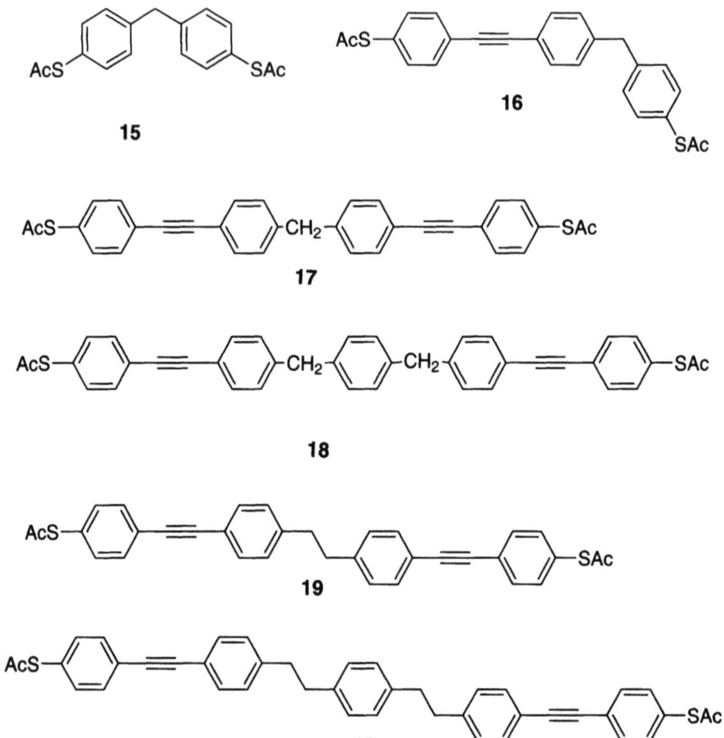

Fig. 4 Molecular wires 15–20, each containing an interior methylene or ethylene group barrier to conduction [38]

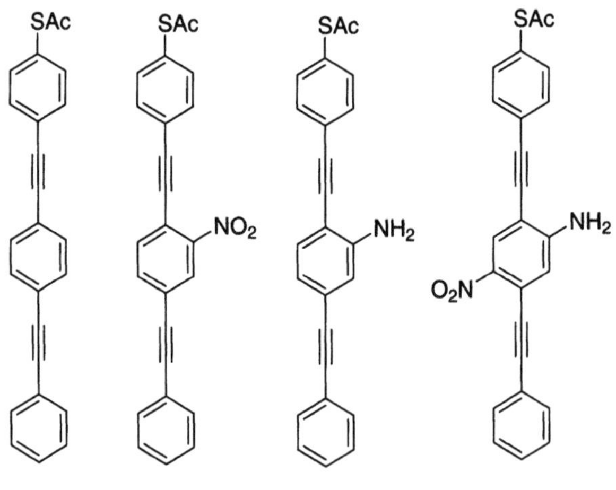

21 22 23 24

Fig. 5 Unfunctionalized OPE 21 and functionalized OPEs 22–24 [26, 41]

lecular wire **21** and amine-functionalized **23** had no activity while nitro-functionalized **22** and **24** were both active switches; conversely, analysis by STM [41] indicated that all three of **21**, **22**, and **24** underwent conformational-based switching (**23** was not tested in the STM experiment) (see [1,2] for more information on switching). This underscores that fact that varying testbeds can afford widely different results, as we established in our review [29].

To further elucidate proposed switching mechanisms of molecular wires, we synthesized U-shaped OPE-based molecules **25–30** (Fig. 6) [43]. The resulting monolayers were studied and the integrity of the molecule–gold attachment was corroborated by single-wavelength ellipsometry (SWE), cyclic voltammetry (CV) and X-ray photoelectron spectroscopy (XPS). Additionally, the kinetics of SAM formation on a gold electrode were probed by CV, while a clear signal corresponding to the thiol–gold bond was detected by XPS. These conformationally restricted oligomers are designed to be of use in studies utilizing scanning probe microscopy techniques to elucidate switching

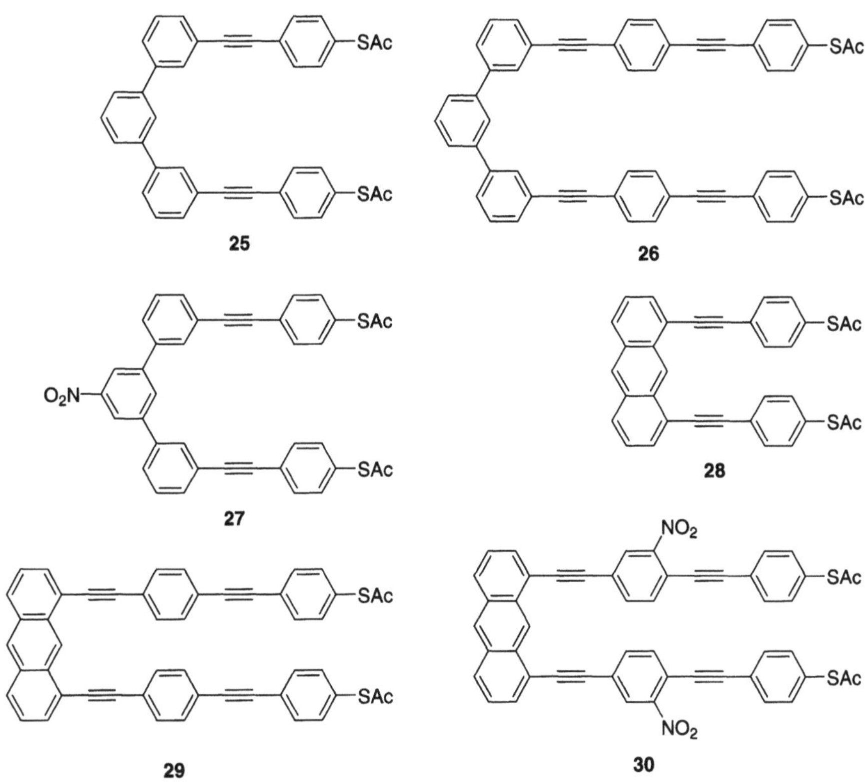

Fig. 6 U-shaped molecules **25–30** synthesized in order to probe the viability of proposed switching mechanisms [43]

mechanisms and negative differential resistance (NDR) behavior thought to be based on molecular conformational changes.

New electron-deficient fluorinated OPEs **31–39** with varied functional groups were synthesized as free thiols, nitriles and pyridines, ready to be used for surface adhesion, see Fig. 7 [44]. Calculated dipole moments suggest better matching between energy levels of bulk interfaces and molecular frontier orbitals when compared to non-fluorinated OPEs. Differential scanning calorimetry confirmed a higher thermal stability than the non-fluorinated counterparts. Surface analysis by ellipsometry, contact angle goniometry, CV, surface infrared and XPS verified that the fluorinated OPEs chemisorb on Au and Pt surfaces. Based on the physical properties of the fluorinated OPEs, they might be useful in future physical vapor deposition techniques, methods that are typically used in standard semiconductor fabrication processes.

We have also synthesized orthogonally functionalized OPEs **40–44**, see Fig. 8 [45]. We have designed these compounds to allow for self-assembly of the molecules via the free thiol or nitrile while protecting the other sulfur atom as a thioacetate to ensure molecular directionality and to inhibit crosslinking if SAM assembly on nanorods is desired. Following initial assembly, the acetate can be removed with NH_4OH or acid to afford the thiolate, which can be assembled onto another metallic material. For the mononitro compounds **40, 43,** and **44**, this affords a monolayer with all the nitro groups

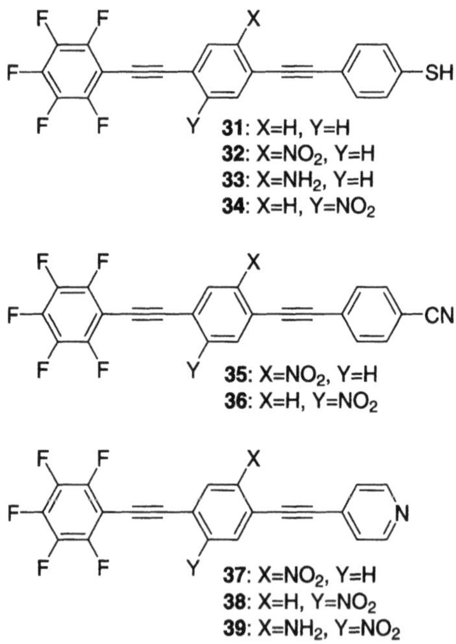

Fig. 7 Fluorinated OPEs **31–39** synthesized for use in physical vapor deposition studies [44]

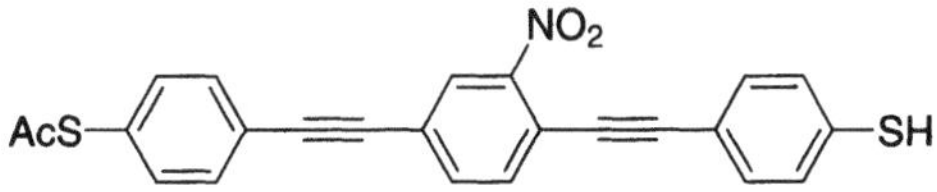

40

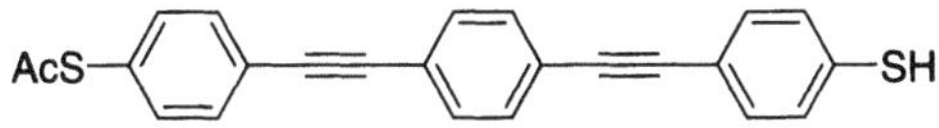

41

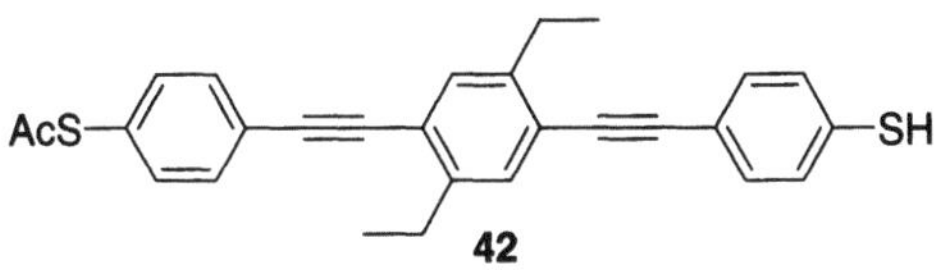

42

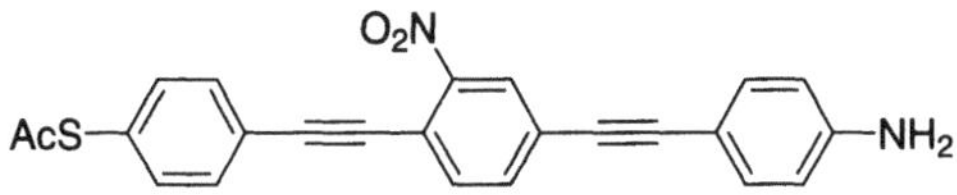

43

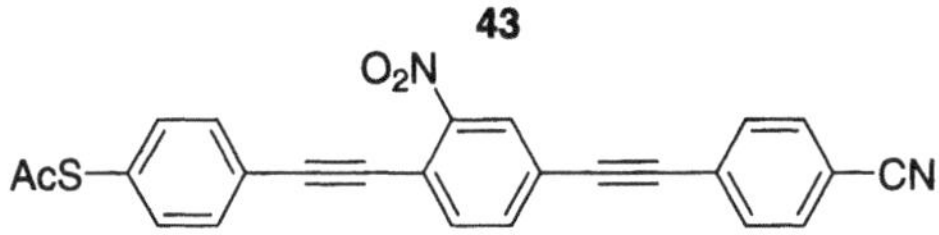

44

Fig. 8 Synthesis of orthogonally functionalized OPEs **40–44** [45]

oriented in the same direction, a result made possible only by the orthogonal protection scheme outlined here.

Martin, Guldi, and coworkers [46] synthesized OPE molecular wires with C_{60} "stoppers" on the termini and found there was no electronic communication between the OPE portion of the molecules and the C_{60} moieties. Our work with OPEs capped by multiple C_{60} molecules came to a similar conclusion, with little interaction between the C_{60} cages and the OPE backbone in the ground state [47].

3.1.3
Oligo(1,4-phenylene vinylene)s

A third class of compounds that has been studied in our labs as well as other's are the oligo(1,4-phenylene vinylene)s (OPVs) [48–54]. As mentioned earlier, the majority of other's work on OPVs has been geared toward applications in the optical and OLED field [49–52, 54]. Our work produced molecular wires targeted for molecular electronics applications, and involved synthesis of the three OPVs **45–47**, shown in Fig. 9. Note that **45** is unfunctionalized, containing only a protected thiol alligator clip for later SAM formation. After formation of the SAM, Au or other metal would be deposited under vacuum for formation of the top contact to complete the circuit through the aromatic

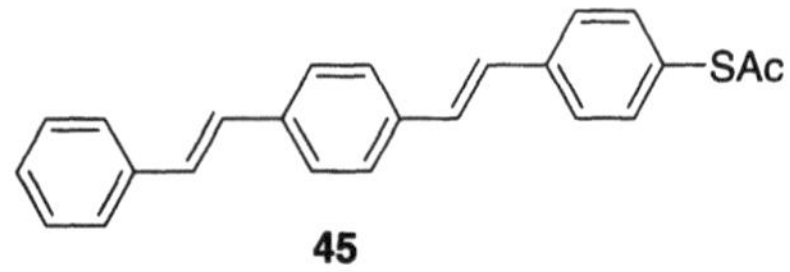

45

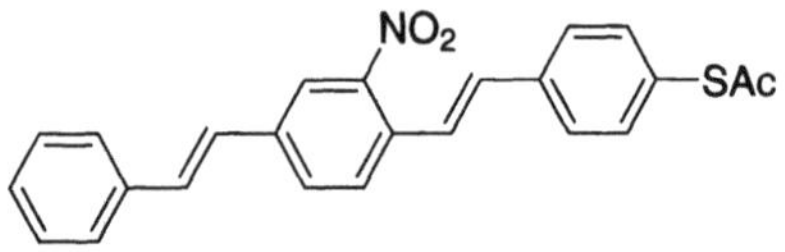

46

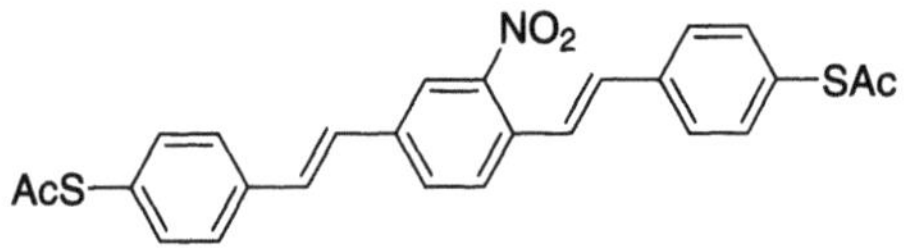

47

Fig. 9 Three oligo(phenylene vinylene)s (OPVs) **45–47** synthesized in order to compare their conductance to OPEs [48]

ring. We functionalized **46** with a nitro group on the interior aromatic core in order to determine if **46** would then act as a switch. Compound **46** is undergoing testing in a collaborator's lab. Finally, **47** was synthesized with both a nitro functional group and protected thiol groups at both ends so that it could span two Au contacts and form a circuit via a self-assembly process. This type of self-assembly process is important in our nanocell research program [2, 55], and is quite different from the approaches of others, who have not necessarily designed OPVs that are meant to connect two proximal probes, an interface to the present solid-state-based technology.

Detert [54] synthesized a series of readily soluble OPVs **48** shown in Fig. 10 to study their electronic spectra. He found that appending various electron-withdrawing groups to the terminal aromatic rings allowed the tuning of the electron affinity of the chromophore without significant changes in the spectra. When those same substituents were placed on the vinylene segments of the OPV molecules, strong bathochromic shifts were observed. Wong investigated similar substituent effects [50]. Meijer [49] and Nierengarten [51] synthesized large assemblies of OPVs attached to other molecules and measured the optical, electronic, and aggregation behaviors of the products.

Chidsey [53] and Wasielewski [52] have used OPVs as core components of molecules (Fig. 11) synthesized to test electron tunneling and long-distance electron transfer, respectively. Of the compounds Chidsey and coworkers synthesized, **49** spanned the most distance, 3.3 nm. The OPVs were assembled on a Au electrode and the tethered ferrocene redox species at the other end of the OPV bridge was exposed to an aqueous electrolyte. They used laser-induced temperature jump techniques to measure the rate constants of thermal, in-

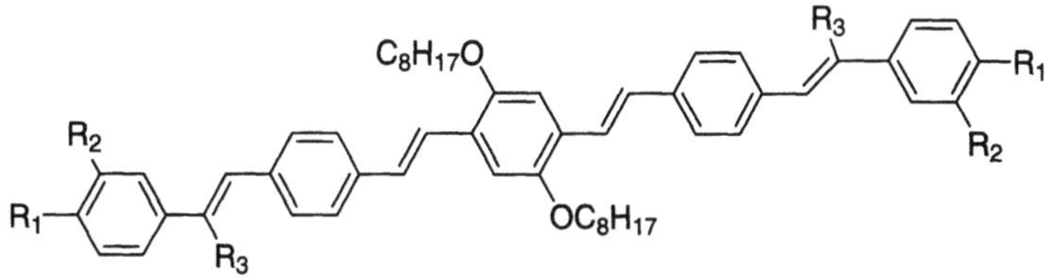

48

$R_1 = CH_3, OC_6H_{13}, Br, CN, NO_2, H, OC_3H_7$, or Cl
$R_2 = H, CF_3, CN, R_4, SO_2C_{10}H_{21}, H$, or CF_3
R_3 = H, or CN

R_4 = 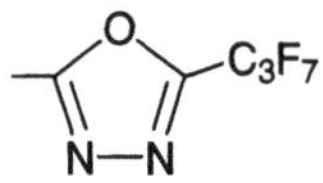

Fig. 10 A series of OPVs **48** synthesized by Detert [54] to study the effect that variation of substituents R_1, R_2, R_3, and R_4 had on the electronic spectra

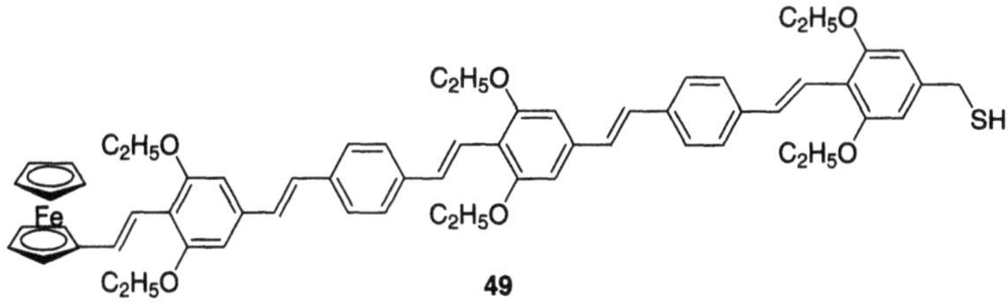

49

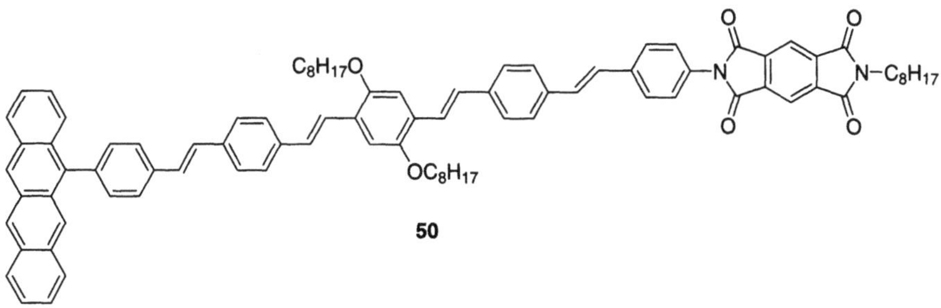

50

Fig. 11 Chidsey [53] synthesized a series of OPVs including **49** to test electron tunneling while Wasielewski [52] synthesized a series of OPVs including **50** to measure long distance electron transfer

terfacial electron transfer through the system and observed transfer through the OPV. They found that OPVs up to 2.8 nm in length were good conductors, with the transfer limited by structural reorganization.

In their study of a series of compounds including **50** (Fig. 11), the lengthiest molecule tested, Wasielewski and coworkers [52] found that electron transfer over long distances depends critically on the low-frequency torsional motions of the molecular wires, i.e. when the molecules twist and turn their molecular orbitals do not line up in a fashion that favors fast electron transfer. But their tests were in solution rather than the more device-realistic molecule-surface attached patterns.

Martin and Guldi studied a homologous series of molecular wires wherein an OPV core connected a π-extended tetrathiafulvalene as an electron donor on one end to a C_{60} as an electron acceptor on the other end [56]. There was strong electronic coupling between the two ends of the molecules, with a donor-acceptor coupling constant (V) of $\sim 5.5\,\text{cm}^{-1}$.

3.1.4 Aromatic Ladder Oligomers

Our group has synthesized a vast array of aromatic ladder polymers [57, 58] for use in conducting polymer and optoelectronic applications. However, we realized that for molecular electronic applications, the molecules we synthesized needed to have defined length and composition in order to be commercially useful as molecular wires.

As shown in Fig. 12, Gourdon and coworkers have synthesized two classes of conjugated ladder oligomers [59] that maintain similar guidelines for molecular wires, including a defined length, rigidity, extended π-conjugation for good electron transfer, and good electronic coupling with metallic con-

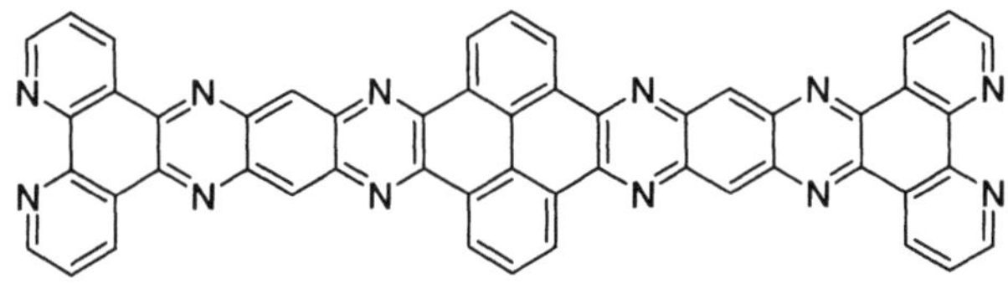

51

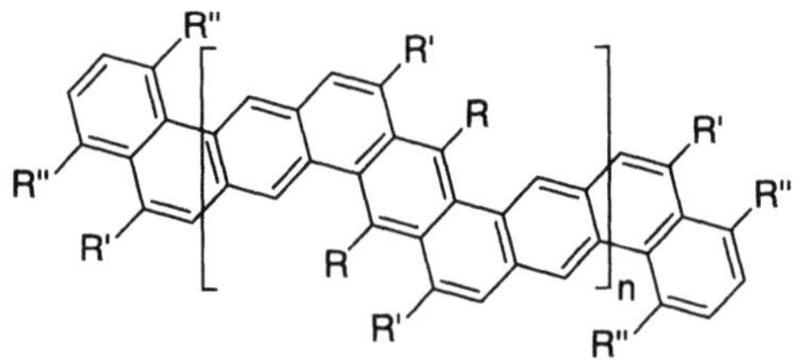

52

$R = C_6H_{13}$
$R' = PhOC_{12}H_{25}$
$R'' = CH_3$

Fig. 12 Gourdon [59] synthesized the aromatic ladder oligomers oligo(quinoxaline) **51** and the oligo(benzoanthracene)s **52**

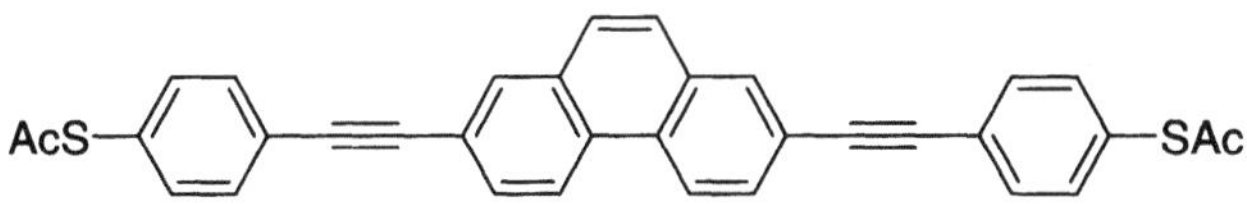

53

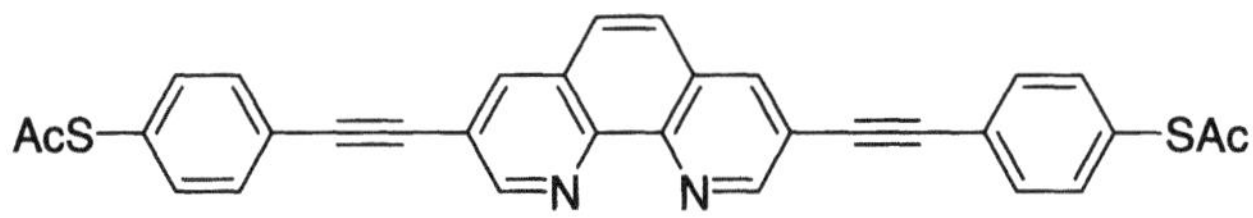

54

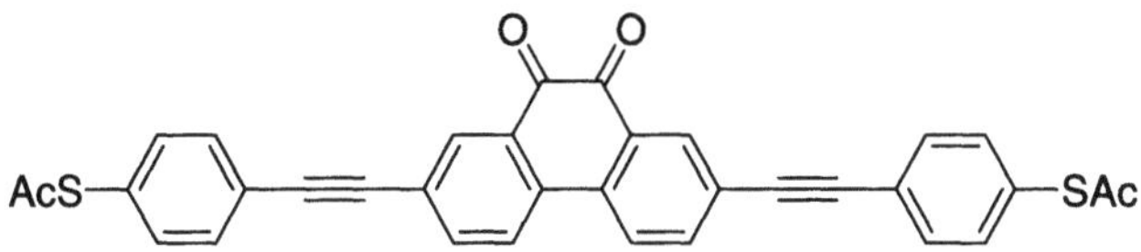

55

Fig. 13 Ladder polyaromatics 53–55 synthesized by our group to study proposed molecular electronic switching mechanisms [61]

tacts. The oligo(quinoxaline) derivatives **51** and oligo(benzoanthracene) molecular wires **52** were synthesized using standard condensation and coupling reactions. The oligo(quinoxaline)s **51** have built-in alligator clips in the terminal 1,10-phenanthroline moieties. We have also synthesized molecular wires containing terminal pyridine and other nitrogen-containing functional groups [39]. Calculations have supported their use as alligator clips [60]. An iterative crossed divergent-convergent process that leads to rapid growth of the oligomers was developed to synthesize the oligo(benzoanthracene)s **52**.

We recently synthesized three molecular systems **53–55** that have ladder-like cores that cannot undergo internal rotational motion (Fig. 13) [61]. Testing of SAMs of shortened versions of the molecules (2-thioacetophenanthrene and 4-thioacetobiphenyl) via STM indicated that internal ring rotation was not required for conductance switching [62].

3.1.5 Oligophenylenes and Polyphenylenes

The oligophenylene and polyphenylene classes of molecules, possessing continuous overlap of molecular orbitals through extended conjugation without intervening groups such as alkynes or olefins, has been an active area of research for our group [63–65]. Müllen and coworkers [66] have synthesized a series of 1-dimensional polyphenylenes including polyindenofluorenes **56** and poly(9,9-diarylfluorene)s **57**, shown in Fig. 14. The photoluminescence (PL) spectra of the series of polyindenofluorenes **56** all have maxima around 430 nm, making them possible candidates for OLED materials as well

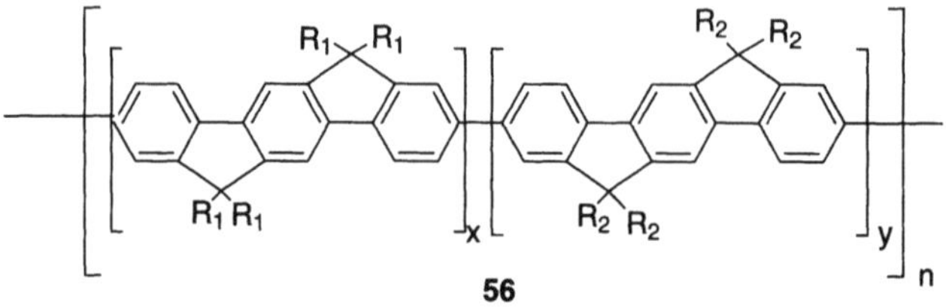

56

R_1 = n=octyl
R_2 = 2-ethylhexyl

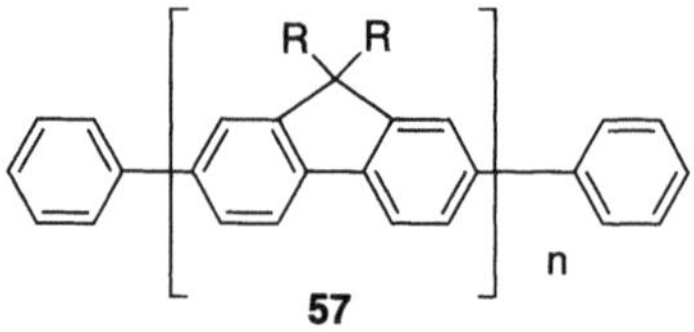

57

R = C_6H_5 or
R = 3,5-di(*tert*-butyl)phenyl or
R = 4-(2',3',4',5'-tetraphenylbiphenyl)

Fig. 14 1-dimensional polyphenylene polymers **56** and **57** that have been synthesized by Müllen [66]

as molecular wires, although for applications in molecular electronics one would normally want to be able to synthesize molecular wires with known lengths and constitutions due to concerns about homogeneity and materials handling. Bryce and coworkers have synthesized molecular wires containing both OPE and fluorenone units, bearing terminal thiol groups [67]. These have been shown to have reversible cathodic solution electrochemistry arising from reduction of the fluorenone units. We have synthesized many oligophenylene derivatives [38, 39] and generally find that the torsional twisting caused by the steric hindrance between the hydrogen atoms at the ortho positions of adjacent phenyl rings leads to decreased orbital overlap, and possibly lower conductance. However, this torsional twisting could be important in improving the switching characteristics of the molecules, so we have recently made terphenyl derivatives **58–63** [68], shown in Fig. 15. The molecules containing mono- and dinitro terphenyl cores were rationally designed based on the electronic properties found in OPEs. The cores were functionalized with nitro groups and with different alligator clips to provide new compounds for testing in the nanopore and planar testbed structures.

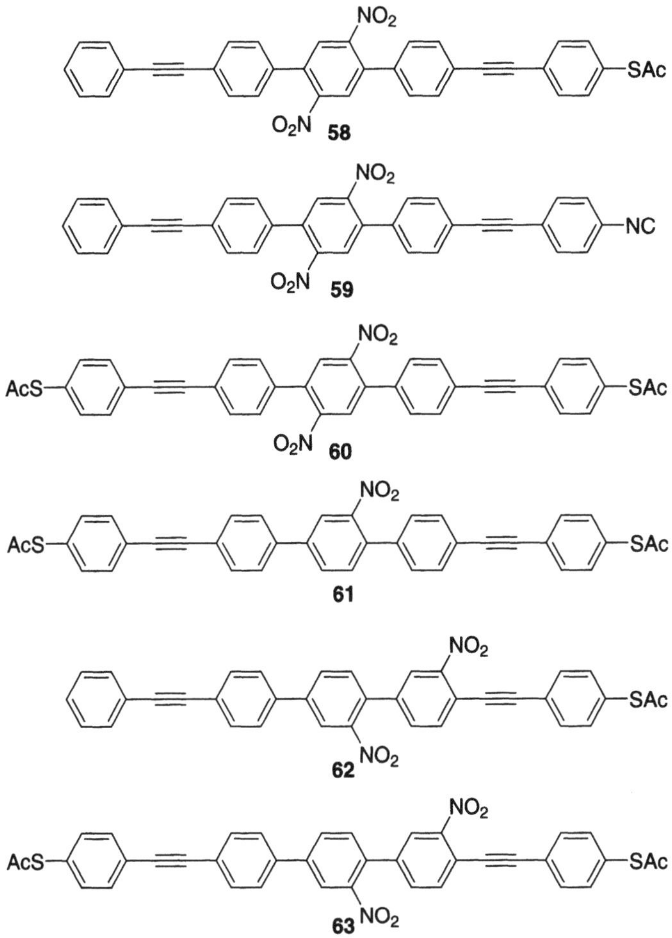

Fig. 15 Terphenyl derivatives **58–63** that have been synthesized to probe proposed switching mechanisms [68]

3.1.6 Acetylene Oligomers

Diederich and coworkers [69, 70] have synthesized the acetylene oligomers class of organic molecular wires. Their work included producing acetylene oligomers that are insulated via dendritic encapsulation, see Fig. 16. They found that the insulated products **64** underwent ready isomerization around the double bond, producing a mixture of *E* and *Z* isomers that made purification of the materials difficult. Oligomeric acetylenic molecular wires that have been encapsulated within zeolites and other mesoporous materials show high electrochemical charge uptake [71]. Anderson and coworkers have published an approach to insulated molecular wires of the oligo(phenylene) class [72].

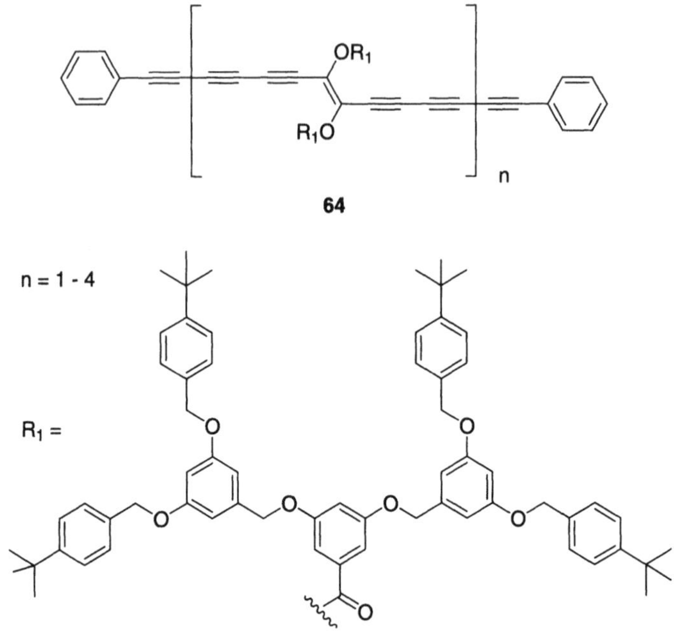

Fig. 16 Encapsulated oligo(pentaacetylene) molecular wires **64** synthesized by Diederich and coworkers [69]

3.1.7 Carbon Nanotubes

Carbon nanotubes have been attractive candidates for use as molecular wires [73, 74]. The so-called "cross-bar" approach to the development of a molecular electronics-based computer has as one approach the use of carbon nanotubes for the wiring between the molecular switches [5, 8] of the circuitry. It is unfortunately difficult to work with carbon nanotubes due to their insolubility in most organic solvents [75], and their tendency to form bundles of tubes that are difficult to separate. Several methods for functionalization of the carbon nanotubes have been developed [76–79] that may make it easier to handle the carbon nanotubes, however the same functionalization techniques can also destroy the electrical properties of the molecules [2]. Calculations by Seifert and coworkers indicate that sidewall fluorination of carbon nanotubes could produce products with a wide range of characteristics from insulating to metallic-like behavior [80]. It is possible that nanowires [21] will be used in the cross-bar computing devices instead of carbon nanotubes due to the easier synthesis and handling of the nanowires.

3.2
Organometallic Molecular Wires

We have synthesized molecules in the porphyrin class of molecular wires [38]. Figure 17 shows four porphyrin derivatives **65–68** made in our labs. Zn, Cu, and Co were all inserted into **65** using the corresponding hydrated metal acetates. Deprotection of the thiol acetates by NH_4OH resulted in no loss of metal, as indicated by subsequent NMR analysis. While being interesting compounds in and of themselves, **67** and **68** could also be intermediates in the synthesis of more complex molecular wires. A four terminal porphyrin **69** shown in Fig. 18 was also synthesized. We have also synthesized [81] and tested the bis(2,5-di-[2]pyridyl-3,4-dithiocyanto-pyrrolate)cobalt(II) complex for its molecular electronics properties and found that in single-molecule transistors made using this Co complex, we observed inelastic cotunneling features that correspond energetically to vibrational excitations of the molecule, as determined by Raman and infrared spectroscopy [82]. This is a form of inelastic electron tunneling spectroscopy of single molecules, with the transistor geometry allowing in-situ tuning of the electronic states via a gate electrode. The vibrational features shift and change shape as the electronic levels are tuned near resonance, indicating significant modification of the vibrational states. When the molecule contains an unpaired electron, we also observe vibrational satellite features around the Kondo resonance.

Lindsey has published a large body of work concerning his synthesis and testing of molecular "photonic" wires based on porphyrin molecules linked by diaryl ethyne units to light absorbing dyes [83]. The energy absorbed by the dyes is transmitted through the porphyrin-diaryl-ethyne wires to a free-base porphyrin transmission unit. The quantum efficiency was determined to be very high, from 81% to > 99%.

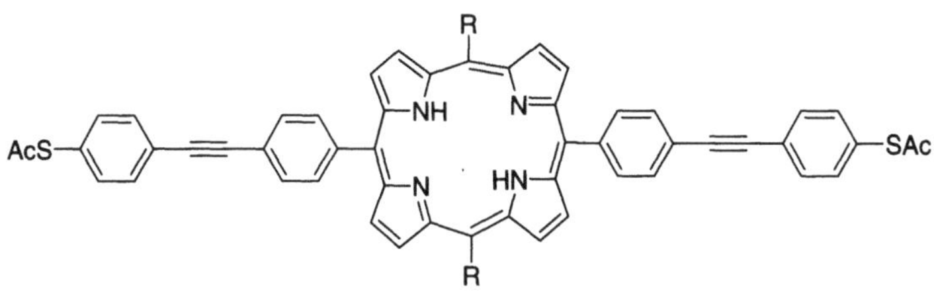

65, R = C_6H_5
66, R = p-C_6H_4-CH_3
67, R = p-C_6H_4-Br
68, R = p-C_6H_4-I

Fig. 17 Porphyrins **65–68** synthesized by us [38]. Zn, Cu, and Co metal atoms were inserted into **65**. Subsequent deprotection of the thiol acetates resulted in no metal ion loss

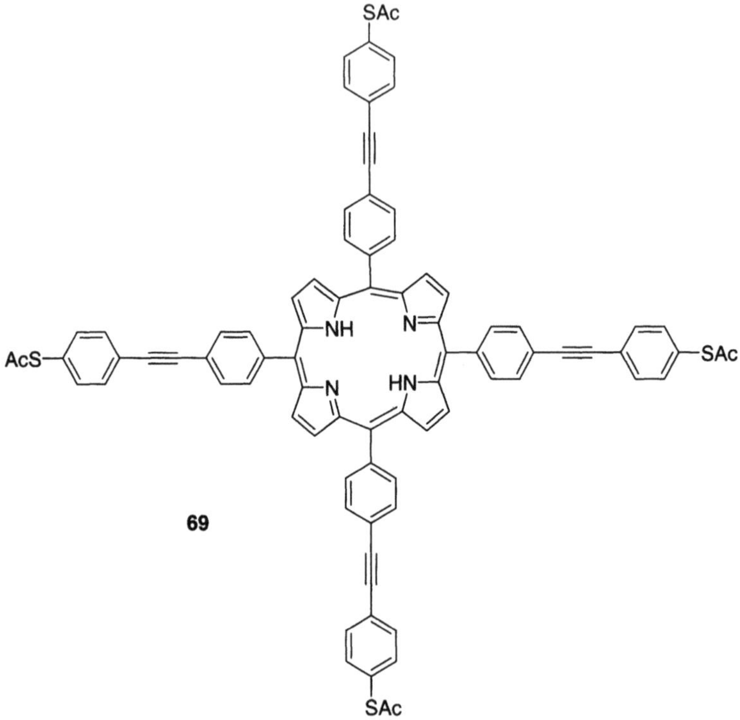

69

Fig. 18 A four terminal porphyrin-based molecular wire **69** synthesized by us [38]

Anderson has reviewed the synthesis and optoelectronic properties of conjugated porphyrin molecular wires [84]. As with the OPTs, OPEs, and oligomeric acetylenes, alkyne moieties have been used to link porphyrin units to make longer molecular wires, such as **70** in Fig. 19. The estimated length of **70** is 8.3 nm from Si atom to Si atom. Anderson's work has shown that the electronic behavior of these types of systems can be attributed to strong inter-porphyrin conjugation in the ground state. This strong interac-

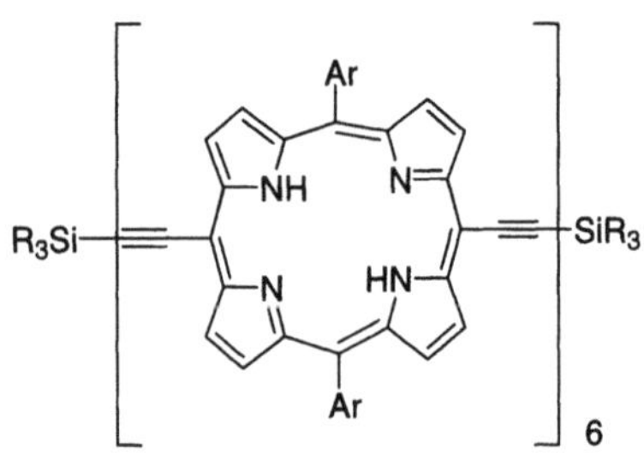

70

Fig. 19 The porphyrin hexamer **70** synthesized by Anderson [84], with six porphyrin units linked through dialkynes

tion is amplified in the excited states, and also in the oxidized and reduced forms of the porphyrin core.

Alessio and coworkers have used self-assembly techniques to synthesize large metallacycles of porphyrins [85]. By synthesizing porphyrins having two peripheral pyridine appendages at either 90° or 180° to each other, and adding the requisite ionic metallic component such as the $RuCl_2$ complex of dimethylsulfoxide (DMSO), supramolecular structures are produced. This self-assembly process proceeds with two porphyrin and two $RuCl_2$ complexes forming (as one example) the molecule **71** shown in Fig. 20. Substitution of CO for DMSO in the $RuCl_2$ raw material forms a similar metallacycle with CO instead of DMSO as Ru ligands. By inserting Zn into the resulting porphyrin core, and exposing the mixture to 4,4′-bipyridine (which acts as a ligand for the Zn atoms at the porphyrin cores), a stacked complex was formed. The authors envisioned that such complexes, with their extended conjugation and metal centers, could harvest light energy and act as molecular wires.

The self-assembly of inorganic molecular wires in solution has been described by Kimizuku [86], however, it is unknown whether this process would produce products usable in constructing devices.

Pyridine ligands are common in organometallic molecular wires [87–89], with a review recently appearing [90]. For instance, Constable and coworkers [87] synthesized the molecular wire **72** shown in Fig. 21. The $[Ru(terpy)_2]^{2+}$ salts by themselves are non-luminescent while adding the 2,5-thiophenediyl spacers produces a molecular wire that is luminescent. Dong has made simi-

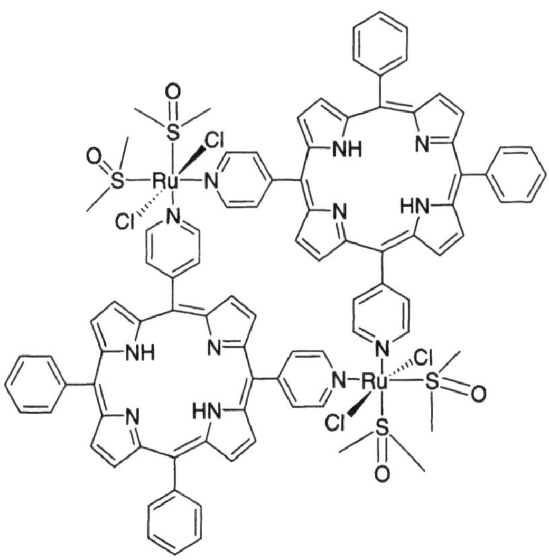

71

Fig. 20 The porphyrin metallacycle 71 synthesized by Allessio [85]

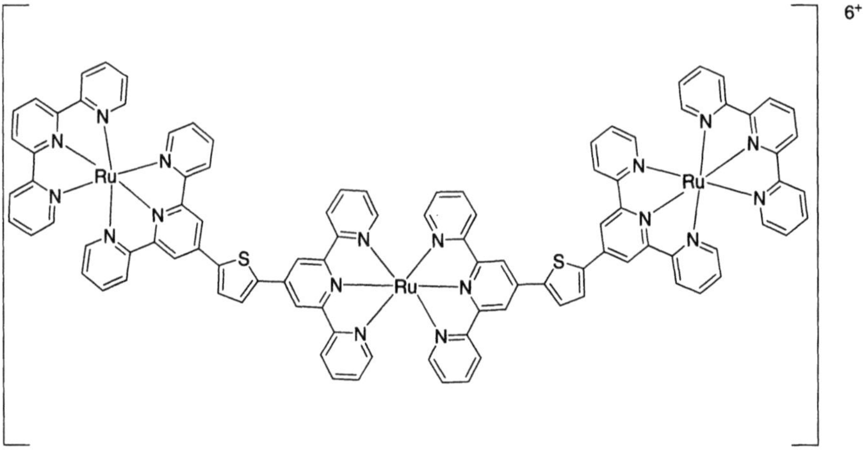

72

Fig. 21 The molecular wire 72 synthesized by Constable [87] that showed luminescence due to the presence of the 2,5-thiophenediyl spacers

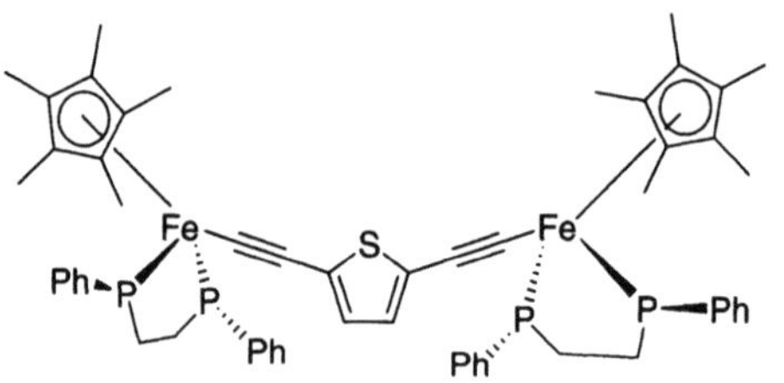

73

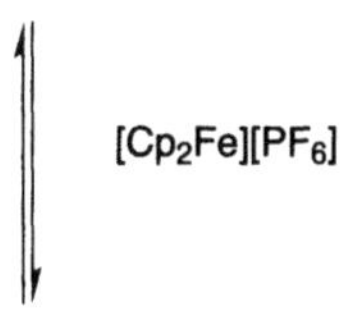

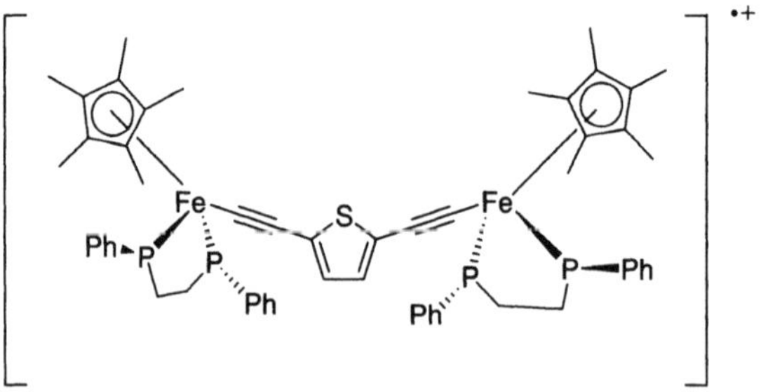

74

Fig. 22 A mixed-valence complex pair of molecular wires 73 and 74 as synthesized by Lapinte [98]

lar Ru complexes with rigid rod spacers [91] while Coudret and Launay have made Ru complexes containing both 2,2′-bipyridine and 2-phenylpyridine ligands, with the metal centers linked by rigid rod cores through the 2-phenylpyridine ligand [92, 93]. Shiotsuka [88] synthesized a Ru – Au – Ru triad by using a bis-σ-Au-acetylide to connect two Ru complexes. The resulting molecular wire showed an intense emission at 620 nm upon excitation at 360 nm, suggesting some energy transfer from the Au to the Ru site via the π-conjugation offered by the ethynyl units.

Cotton [89] synthesized a trinickel complex of the ligand di-2,2′-pyridylamide that had a deep blue color and showed metal–metal bonding interaction in the crystal structure. This work has been extended to include complexes containing $^{-}$CN, $^{-}$NCNCN, and $^{-}$C≡CPh [94], the addition of substituents to the di-2,2′-pyridylamide ligands to increase the stability of the trinickel complexes [95], and addition of polyalkynyl assemblies to the complexes where the metal is Cr, Co, or Ni [96]. The polymeric organometallic molecular wire $[Os(0)(2,2'\text{-bipyridine})(CO)_2]_2$ has been formed from a mononuclear Os(II) carbonyl precursor and found to be soluble in butyronitrile [97].

An interesting mixed-valence molecular wire pair **73** and **74** synthesized by Lapinte [98] is shown in Fig. 22. As we have seen in many of the molecular wires, both the thiophenyl and the alkynyl units are present in this molecular wire.

4
Measurement of Conduction in Molecular Wires

It is difficult to discuss molecular wires without covering their use in device embodiments or testbeds. A large body of work has been published concerning the measurement of conductance in molecular wires [99–110]. We will touch on the highlights in this chapter; readers wishing to have more in-depth information are invited to see our recent review [29].

Reed has studied the conductance of both single molecules [25] and of SAMs of molecules formed in a nanopore device [101]. Stable and reproducible switching and memory effects were been seen in the nanopore devices, with demonstrated NDR and charge storage with bit retention times of greater than 15 min at room temperature.

Bourgoin [103] compared the electronic coupling efficiency of S and Se alligator clips on Au surfaces and found Se to be the better coupling link. Whitesides has developed a testbed for measuring conductance in molecular wires [105] using two metal electrodes sandwiching two SAMs, with the top metal electrode being Hg for convenience of formation. There are three different forms of this testbed; the first comprises two Hg drops, each covered with the same SAM; the second testbed comprises a Hg drop covered with a SAM interacting with a SAM formed on Ag; the third testbed is similar to the first, only with redox-active molecules trapped between the two SAMs in order to do electrochemical measurements on them.

Avouris [106] has carried out extensive conductance measurements on carbon nanotubes and has constructed electronic devices containing them, including carbon nanotube field-effect transistors (CNTFETs). Nanotubes are known to have either metallic or semiconducting properties. One problem with single-walled nanotube (SWNT) bundles is separating the metallic tubes from the semiconducting tubes. Avouris has developed a process that removes the metallic nanotubes, leaving the semiconducting behind.

Kushmerick and coworkers [108] have developed a testbed device in which a SAM of the molecule of interest is formed on one of two 10 μm Au wires that are crossed, and brought into contact by the Lorentz force, i.e. DC current in one wire deflects it in a magnetic field. Using this device, $I(V)$ characteristics were measured for three molecules **75–77** shown in Fig. 23. Normalizing the conductance of the C_{12} alkanethiol **75** to 1, the conductance of the OPE molecular wire **76** was measured at 15 and the conductance of the OPV molecular wire **77** was measured at 46. Note that **77** has benzylic thioester moieties with methylene barrier groups to disrupt the molecular orbital overlap of the rest of the conjugated system. The researchers attribute the increased conductivity of the OPV molecule to both the increased coplanarity of the molecule (the alkynes in the OPEs allow the phenyl rings to be more freely rotating

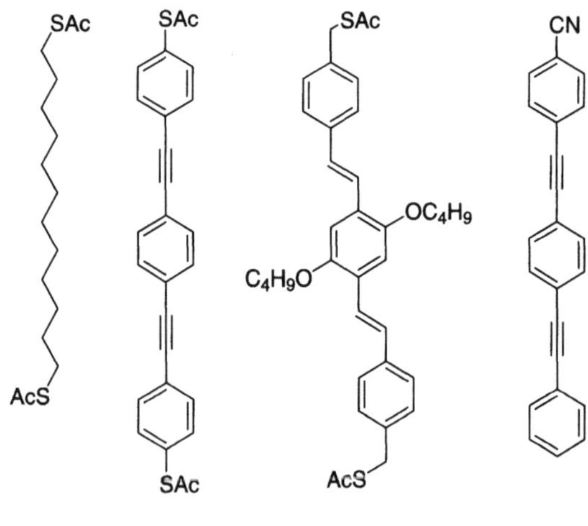

75 **76** **77** **78**

Fig. 23 The molecular wires **75–77** tested by Kushmerick and coworkers [108] using a crossed-wire tunnel junction testbed to determine their relative conductance. The OPEs **76** and **78** and the OPV **77** were tested by Kushmerick [111] and Blum [112] both in the crossed-wire testbed and using STM

when compared to the vinylene group-linked OPVs, thus the population of OPE molecules existing in fully conjugated form is lower than the population of OPV molecules existing in fully conjugated form at the same temperature), and to the more regular periodicity of the conjugated molecular backbone of the OPV molecular wire. Compare the short 0.1218 nm alkyne linkage in the OPE molecular wire to the 0.1352 nm vinylene linkage of the OPV molecular wire and the 0.141 ± 0.001 nm periodicity of the π-conjugated molecular backbone. Kushmerick [111] and Blum [112] found that the conductances of OPEs **76** and **78** and OPV **77** shown in Fig. 23 scaled according to the number of molecules in the crossed-wire junction and were about 10^3 times higher than that for single molecules isolated in a C11 alkane thiolate matrix and tested by STM measurements [113].

Dholakia and coworkers [114] have shown that the $I(V)$ characteristics of SAMs formed from **21** (Fig. 5) depend on orientation, packing, and order of the SAM. They used STM, ellipsometry, and XPS in their analyses of the SAMs.

Rawlett and coworkers [109] used conducting atomic force microscopy (cAFM) to measure the conductance of two different molecules that had been inserted into the naturally occurring defect sites of a dodecanethiol SAM on Au. The molecular wires **76** (Fig. 23) and **79** (Fig. 24) were constrained to be standing upright, parallel to the surface of the Au, due to the tight packing of the dodecanethiol SAM. Au nanoparticles were attached to the projecting deprotected thiol groups, and the nanoparticle was contacted by a Au-coated AFM tip to measure the conductance through what are thought to be indi-

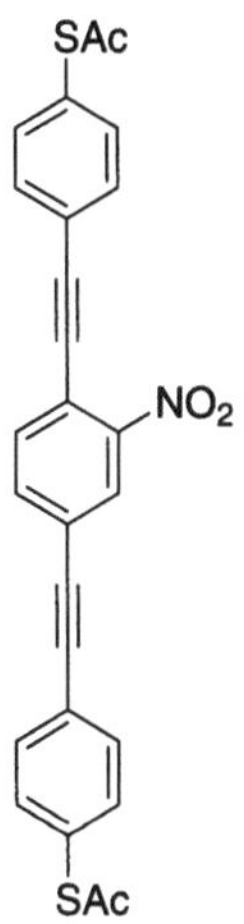

79

Fig. 24 The molecular wire **79**, the conductance of which was tested by Rawlett and coworkers [109] using cAFM

vidual molecules based on evidence not discussed here. Reproducible NDR effects were seen only for molecular wire **79**, not for **76**, indicating the functionalization of the wire with the $-NO_2$ group leads to the NDR effect.

Ishida and coworkers [115] used cAFM to measure the conductance of conjugated molecules including benzenethiol, benzylthiol, and biphenyl and terphenyl thiol derivatives as well as OPVs including **45** (Fig. 9) and found that OPVs had higher conductances than the other molecules. They observed apparent NDR behavior at higher SAMs bias measurements; the appearance of the NDR might be related to the roughness of the SAM surface.

Reed has completed a systematic temperature-dependent study of the conductance of alkanethiolates [110] in SAMs and has shown conclusively that direct tunneling is the dominant transport mechanism.

5 Conclusion

Based on the wealth of literature cited, fully conjugated completely organic aromatic molecular wires are the best candidates for introduction into new electronic devices as replacements for the Al or Cu wiring presently used in logic and memory devices. The OPE and OPV classes of molecular wires with S alligator clips have the highest conductances, both theoretical [9] and measured. Recent data indicates that OPV molecular wires are better than OPE molecular wires. Many more questions remain to be answered, the most important of which is how these molecular wires will be integrated into processes to build the molecular electronic and optoelectronic devices of the future. Problems raised by such integration efforts will likely require several iterations in molecular wire research. As we have pointed out, the beauty of organic chemistry is that simple changes to the raw materials used in synthesizing the molecular wires can yield products with vastly different physical properties. The molecular wire(s) that eventually appears in commercial devices may bear no resemblance to those we have discussed. Thus research in this area can still yield much fruit.

Acknowledgements We thank DARPA administered by the Office of Naval Research (ONR); the ONR Polymer Program; the US Department of Commerce, National Institute of Standards and Technology; the Army Research Office; NASA; and the Air Force Office of Scientific Research (AFOSR F49620-01-1-0364) for financial support of this work.

References

1. Tour JM (2003) Molecular electronics: commercial insights, chemistry, devices, architecture, and programming. World Scientific Publishing, River Edge, New Jersey

2. Tour JM, James DK (2002) Molecular electronic computing architecture. In: Goddard WA, Brenner DW, Lyshevski SE, Iafrate GJ (eds) Handbook of nanoscience, engineering, and technology. CRC Press, Boca Raton, Florida, p 4:1–4-28
3. Ward MD (2001) J Chem Ed 78:321
4. Tour JM (2000) Acc Chem Res 33:791
5. Heath JR (2000) Pure Appl Chem 72:11
6. Reed MA, Tour JM (2000) Sci Am June:68
7. Overton R (2000) Wired 8:242
8. Heath JR, Kuekes PJ, Snider GS, Williams RS (1998) Science 280:1716
9. James DK, Tour JM (2004) Molecular wires. In: Encyclopedia of nanoscience and nanotechnology. Marcel Dekker, New York, p 2177
10. Robertson N, McGowan CA (2003) Chem Soc Rev 32:96
11. Moore GE (1965) Electronics 38. Available on the web: ftp://download.intel.com/research/silicon/moorespaper.pdf (accessed December 2004)
12. Mutschler AS (2004) Electronic News 13 July
13. Intel press release (2004) http://www.intel.com/pressroom/archive/releases/20040614corp.htm (accessed January 2005)
14. Singer P (2004) Semiconductor International 1 December
15. International Technology Roadmap for Semiconductors web pages (2004) http://www.itrs.net/Common/2004Update/2004Update.htm (accessed January 2005)
16. Wang KL (2002) J Nanosci Nanotech 2:235
17. Patel NK, Cinà S, Burroughes JH (2002) IEEE J Sel Topics in Quan Electronics 8:346
18. Popovic ZD, Aziz H (2002) IEEE J Sel Topics in Quan Electronics 8:362
19. Matthews SJ (2001) Laser Focus World 37:169
20. Cotton FA, Daniels LM, Murillo CM, Wang X (1999) Chem Commun 1999:2461
21. Hu J, Odom TW, Lieber CM (1999) Acc Chem Res 32:435
22. Chung S-W, Yu J-Y, Heath JR (2000) Appl Phys Lett 76:2068
23. Cui Y, Lieber CM (2001) Science 291:851
24. Gudiksen MS, Wang J, Lieber CM (2001) J Phys Chem B 105:4062
25. Reed MA, Zhou C, Muller CJ, Burgin TP, Tour JM (1997) Science 278:252
26. Chen J, Reed MA, Rawlett AM, Tour JM (1999) Science 286:1550
27. Ranganathan S, Steidel I, Anariba F, McCreery RL (2001) Nano Lett 1:491
28. Fan F-RF, Yang J, Cai L, Price DW, Dirk SM, Kosynkin DV, Yao Y, Rawlett AM, Tour JM, Bard AJ (2002) J Am Chem Soc 124:5550
29. James DK, Tour JM (2004) Chem Mater 16:4423
30. Bumm LA, Arnold JJ, Cygan MT, Dunbar TD, Burgin TP, Jones L, Allara DL, Tour JM, Weiss PS (1996) Science 271:1705
31. Pearson DL, Tour JM (1997) J Org Chem 62:1376
32. Pearson DL, Jones L, Schumm JS, Tour JM (1997) Synthetic Metals 84:303
33. Pearson DL, Jones L, Schumm JS, Tour JM (1997) Synthesis of molecular scale wires and alligator clips. In: Joachim C, Roth S (eds) Proc NATO advanced research workshop on atomic and molecular wires, Les Houches, France, 6–10 May, 1996. Applied Science Series E, Kluwer Academic Publishers, Dordrecht 341:81
34. Tour JM (2000) Polymer News 25:329
35. Jones L, Schumm JS, Tour JM (1997) J Org Chem 62:1388
36. Collman JP, Zhong M, Constanzo S, Sunderland CJ, Aukauloo A, Berg K, Zeng L (2001) Synthesis 2001:367
37. Gu T, Nierengarten J-S (2001) Tet Lett 42:3175
38. Tour JM, Rawlett AM, Kozaki M, Yao Y, Jagessar RC, Dirk SM, Price DW, Reed MA, Zhou C-W, Chen J, Wang W, Campbell I (2001) Chem Eur J 7:5118

39. Dirk SM, Price DW, Chanteau S, Kosynkin DV, Tour JM (2001) Tetrahedron 57:5109
40. Hwang JJ, Tour JM (2002) Tetrahedron 58:10387
41. Donhauser ZJ, Mantooth BA, Kelly KF, Bumm LA, Monnell JD, Stapleton JJ, Price DW, Rawlett AM, Allara DW, Tour JM, Weiss PS (2001) Science 292:2303
42. Chen J, Wang W, Klemic J, Reed MA, Axelrod BW, Kaschak DB, Rawlett AM, Price DW, Dirk SM, Tour JM Grubisha DS, Bennett DW (2002) Ann N Y Acad Sci 960:69
43. Maya F, Flatt AK, Stewart MP, Shen DE, Tour JM (2004) Chem Mater 16:2987
44. Maya F, Chanteau SH, Cheng L, Stewart MP, Tour JM (2005) Chem Mater 17:1331
45. Flatt AK, Yao Y, Maya F, Tour JM (2004) J Org Chem 69:1752
46. Atienza C, Insuasty B, Seoane C, Martin N, Ramey J, Rahman GMA, Guldi DM (2005) J Mater Chem 15:124
47. Zhao Y, Shirai Y, Slepkov AD, Cheng L, Alemany LB, Sasaki T, Hegmann FA, Tour JM (2005) Chem Eur J 11:3643
48. Flatt AK, Dirk SM, Henderson JC, Shen DE, Su J, Reed MA, Tour JM (2003) Tetrahedron 59:8555
49. Syamakumari A, Schenning APHJ, Meijer EW (2002) Chem Eur J 8:3353
50. Wong MS, Li ZH, Shek MF, Samroc M, Samoc A, Luther-Davies B (2002) Chem Mater 14:2999
51. Gu T, Ceroni P, Marconi G, Armaroli N, Nierengarten J-F (2001) J Org Chem 66:6432
52. Davis WB, Ratner MA, Wasielewski MR (2001) J Am Chem Soc 123:7877
53. Sikes HD, Smalley JF, Dudek SP, Cook AR, Newton MD, Chidsey CE, Feldberg SW (2001) Science 291:1519
54. Detert H, Sugiono E (2000) Synth Metals 115:89
55. Tour JM, Cheng L, Nackashi DP, Yao Y, Flatt AK, St Angelo SK, Mallouk TE, Franzon PD (2003) J Am Chem Soc 125:13279
56. Giacalone F, Segura JL, Martin N, Guldi DM (2004) J Am Chem Soc 126:5340
57. Zhang CY, Tour JM (1999) J Am Chem Soc 121:8783
58. Yao Y, Tour JM (1999) Macromolecules 32:2455
59. Gourdon A (1997) Synthesis of conjugated ladder oligomers. In: Joachim C, Roth S (eds) Proc NATO advanced research workshop on atomic and molecular wires, Les Houches, France, 6–10 May, 1996. Applied Science Series E, Kluwer Academic Publishers, Dordrecht 341:81
60. Billić A, Reimers JR, Hush NS (2002) J Phys Chem B 106:6740
61. Ciszek JW, Tour JM (2004) Tetrahedron Lett 45:2801
62. Dameron AA, Ciszek JW, Tour JM, Weiss PS (2004) J Phys Chem B 108:16761
63. Tour JM (1994) Adv Mater 6:190
64. Tour JM, John JA (1993) Polym Prepr (Am Chem Soc Div Polym Chem) 34:372
65. Tour JM, Lamba JJS (1993) J Am Chem Soc 115:4935
66. Grimsdale AC, Müllen K (2001) The Chemical Record 1:243
67. Wang C, Batsanov AS, Bryce MR, Sage I (2004) Org Lett 6:2181
68. Maya F, Tour JM (2004) Tetrahedron 60:81
69. Schenning APHJ, Arndt J-D, Ito M, Stoddart A, Schrieber M, Siemsen P, Martin RE, Boudon C, Gisselbrecht JP, Gross M, Gramlich V, Diederich F (2001) Helv Chim Acta 84:296
70. Livingston RC, Cox LR, Gramlich V, Diederich F (2001) Angew Chem Int Ed 40:2334
71. Alvaro M, Ferrer B, García M, Lay A, Trinidad F, Valenciano J (2002) Chem Phys Lett 356:577
72. Taylor PN, O'Connell MJ, McNeill LA, Hall MJ, Aplin RT, Anderson HL (2000) Angew Chem Int Ed 39:3456

73. Kong J, Franklin NR, Zhou C, Chapline MC, Peng S, Cho K, Dai H (2000) Science 287:622
74. Franklin NR, Li Y, Chen RJ, Javey A, Dai H (2001) Appl Phys Lett 79:4571
75. Bahr JL, Mickelson ET, Bronikowski MJ, Smalley RE, Tour JM (2001) Chem Commun 2001:193
76. Bahr JL, Tour JM (2002) J Mater Chem 12:1952
77. Bahr JL, Tour JM (2001) Chem Mater 13:3823
78. Dyke CA, Tour JM (2004) Chem Eur J 10:812
79. Hudson JL, Casavant MJ, Tour JM (2004) J Am Chem Soc 126:11158
80. Seifert G, Köhler T, Frauenheim T (2000) Appl Phys Lett 77:1313
81. Ciszek J, Tour JM Unpublished results
82. Yu LH, Keane ZK, Ciszek JW, Cheng L, Stewart MP, Tour JM, Natelson D (2004) Phys Rev Lett 92:266802
83. Ambroise A, Kirmaier C, Wagner RW, Loewe RS, Bocian DF, Holten D, Lindsey JS (2002) J Org Chem 76:3811 and references cited therein
84. Anderson HLB (1999) Chem Commun 1999:2323
85. Iengo E, Zangrando E, Minatel R, Alessio E (2002) J Am Chem Soc 124:1003
86. Kimizuku N (2000) Adv Mater 12:1461
87. Constable EC, Housecroft CE, Schofield ER, Encinas S, Armaroli N, Barigelletti F, Flamigni L, Figgemeier E, Vos JG (1999) Chem Commun 1999:869
88. Shiotsuka M, Yamamoto Y, Okuno S, Kitou M, Nozaki K, Onaka S (2002) Chem Commun 2002:590
89. Berry JF, Cotton FA, Daniels LM, Murillo CA (2002) J Am Chem Soc 124:3212
90. Barigelletti F, Flamigini L (2000) Chem Soc Rev 29:1
91. Dong T-Y, Lin M-C, Chiang MY-N, Wu J-Y (2004) Organometallics 23:3921
92. Fraysse S, Coudret C, Launay J-P (2003) J Am Chem Soc 125:5880
93. Hortholary C, Coudret C (2003) J Org Chem 68:2167
94. Berry JF, Cotton FA, Murillo CA (2003) Dalton Trans 2003:3015
95. Berry JF, Cotton FA, Lu T, Murillo CA, Wang X (2003) Inorg Chem 42:3595
96. Berry JF, Cotton FA, Murillo CA, Roberts BK (2004) Inorg Chem 43:2277
97. Hartl F, Mahabiersing T, Chardon-Noblat S, Da Costa P, Deronzier A (2004) Inorg Chem 43:7250
98. Stang SL, Paul F, Lapinte C (2000) Organometallics 19:1035
99. Lehmann J, Kohler S, Hänggi P, Nitzan A (2002) Phys Rev Lett 88:228305-1
100. Weber HB, Reichert J, Weigend F, Ochs R, Beckmann D, Mayer M, Ahlrichs R, Löhneysen H (2002) Chem Phys 281:113
101. Chen J, Reed MA (2002) Chem Phys 281:127
102. Agraït N, Untiedt C, Rubio-Bollinger G, Vieira S (2002) Chem Phys 281:231
103. Patrone L, Palacin S, Bourgoin JP, Laboute J, Zambelli T, Gauthier S (2002) Chem Phys 281:325
104. Davis WB, Ratner MA, Wasielewski MR (2002) Chem Phys 281:333
105. Rampi MA, Whitesides GM (2002) Chem Phys 281:373
106. Avouris P (2002) Chem Phys 281:429-445
107. Cuniberti G, Fagas G, Richter K (2002) Chem Phys 281:465
108. Kushmerick JG, Holt DB, Pollack SK, Ratner MA, Yang JC, Schull TL, Naciri J, Moore MH, Shashidhar R (2002) J Am Chem Soc 124:10654
109. Rawlett AM, Hopson TJ, Nagahara LA, Tsui RK, Ramachandran GK, Lindsay SM (2002) Appl Phys Lett 81:3043
110. Lee T, Wang W, Reed MA (2003) Ann NY Acad Sci 21
111. Kushmerick JG, Naciri J, Yang JC, Shashidar R (2003) Nano Lett 3:897

112. Blum AS, Kushmerick JG, Pollack SK, Yang JC, Moore M, Naciri J, Shashidar R, Ratna BR (2004) J Phys Chem B 108:18124
113. Blum AS, Yang JC, Shashidar R, Ratna B (2003) Appl Phys Lett 82:3322
114. Dholakia GR, Fan W, Koehne J, Han J, Meyyappan M (2004) Phys Chem Rev B 69:153402
115. Ishida T, Mizutani W, Liang T-T, Azehara H, Miyake K, Sasaki S, Tokumoto H (2003) Ann NY Acad Sci 1006:164

Top Curr Chem (2005) 257: 63–102
DOI 10.1007/b136067

Published online: 14 July 2005

Photoinduced Electron/Energy Transfer Across Molecular Bridges in Binuclear Metal Complexes

C. Chiorboli · M. T. Indelli · F. Scandola (✉)

Dipartimento di Chimica dell'Università, 44100 Ferrara, Italy
snf@unife.it

1	**Introduction**	64
2	**Electron Transfer Mechanisms and Bridge Effects**	66
2.1	Superexchange	68
2.2	Charge Hopping	70
2.3	Interplay of Mechanisms	71
3	**Electronic Energy Transfer**	71
4	**Peculiarities of MLCT Chromophores**	74
5	**Case Studies**	77
5.1	Dyads with Poly-*p*-phenylene Bridges	78
5.1.1	Ru(II)–Rh(III) Systems	78
5.1.2	Ru(II)–Os(II) Systems	84
5.2	Dyads with Extended Aza-aromatic Bridges	88
5.2.1	tpphz Complexes	89
5.2.2	tatpp Complexes	95
5.2.3	bqpy Complexes	97
6	**Conclusions**	98
References		99

Abstract Molecular bridges that efficiently move charge between remote donor and acceptor sites can be thought of as molecular wires. Insight into the properties of molecular wires can be obtained by studying photoinduced electron transfer in covalently linked donor–bridge–acceptor systems. This article summarizes some of the recent progress in the study of such systems involving transition metal complexes as donor and acceptor units. Specific classes of molecular bridges are considered, namely, polyphenylene, and polyquinoxaline bridges. Basic questions are discussed, such as the transfer mechanisms, the associated distance and bridge structure dependence, and the interplay between energy and electron transfer.

Keywords Electron transfer · Energy transfer · Supramolecular photochemistry · Bimetallic complexes · Molecular wires

Abbreviations

cr	Charge recombination
cs	Charge separation
ET	Electron transfer
EnT	Energy transfer
FCWD	Franck–Condon weighted density of states
HOMO	Highest occupied molecular orbital
ILET	Intraligand electron transfer
LC	Ligand centered
LMCT	Ligand-to-metal charge transfer
LUMO	Lowest unoccupied molecular orbital
MMET	Metal-to-metal electron transfer

Ligands

bpy	2,2′-Bipyridine
bqpy	Bis-{dipirido[3,2-*f*:2′,3′-*h*]quinoxalo}-[2,3-*e*:2′,3′*l*]pyrene
Me_2bpy	4,4′-Dimethyl-2,2′-bipyridine
Me_2phen	4,7-Dimethyl-1,10-phenanthroline
phen	1,10-Phenanthroline
tatpp	9,11,20,22-Tetraazatetrapyrido[3,2-*a*:2′3′-*c*3″,2″-*l*:2‴,3‴-*n*]pentacene
tpphz	Tetrapyrido[3,2-*a*:2′,3′-c:3″,2″-*h*:2‴,3‴-*j*]phenazine
tpy	2,2,2″-Terpyridine

1 Introduction

According to the paradigm of molecular electronics [1–5], individual molecules or suitable assemblies thereof (supramolecular systems) can be designed to emulate typical functions of electronic devices. The ultimate extension of this concept would be the construction, in a bottom-up approach, of electronic circuits (computers) from molecules.

The simplest component of an electrical circuit is a wire, and the design of molecular wires has received a great deal of attention. In a very broad sense, this term can be used to designate any molecular[1] structure able to mediate the transfer of electrons between appropriate donor and acceptor sites (electrodes, photo- and redox-active molecular components). In practice, different conduction mechanisms may apply, depending on the molecular structure of the wire and on the type of experimental setup used (see below). Molecular wires have been studied in a variety of experimental conditions, depending on the nature of the donor and acceptor terminals the wire is connected to, and on the method used to detect the electron flow. Available methods include: (i) fast electrochemistry of self-assembled monolayers con-

[1] This restricts the field to discrete molecular systems. In a broad sense, the field of "molecular wires" includes, however, some very interesting work more extended systems, such as carbon nanotubes [102, 103] and one-dimensional semiconductors [104].

taining electroactive groups [6–10], (ii) conductance [$I(V)$] measurements on metal–wire–metal junctions [11–19], (iii) photoinduced electron transfer in donor–bridge–acceptor systems [20–23].

Photoinduced electron transfer in a donor(A)-bridge(L)-acceptor(B) system is schematically depicted in Fig. 1, where (as it is the case in most of the examples to be discussed later) light excitation of the donor is assumed. The alternative case involving excitation of the acceptor is also possible (in that case the electron transfer from donor to excited acceptor is often called hole transfer). In this review article, we will discuss photoinduced electron transfer in donor-bridge acceptor-systems (dyads) involving transition metal complexes as donor and acceptor units.

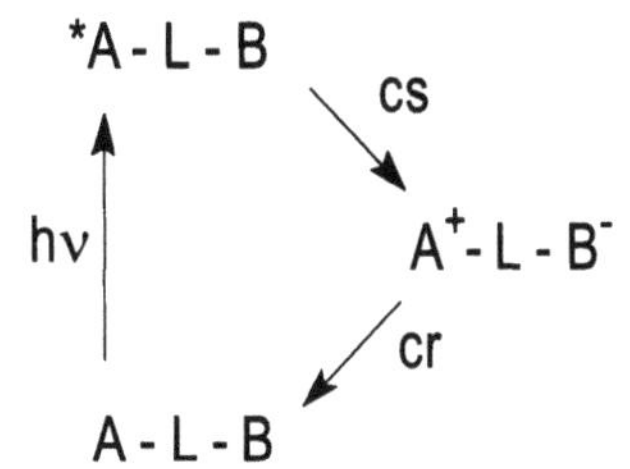

Fig. 1 Schematic representation of photoinduced electron transfer in a donor–bridge–acceptor system, showing forward (charge separation, cs) and back (charge recombination, cr) electron transfer steps

As outlined below (Sect. 3), electronic energy transfer (Fig. 2) is a process that shares with electron transfer a number of formal and substantial similarities. Especially in metal-containing dyads, electronic energy transfer and photoinduced electron transfer are often strongly intertwined. Therefore, the discussion will not be strictly limited to electron transfer processes, but will include, when required, energy transfer results as well. Recent comprehensive reviews on donor–bridge–acceptor systems containing metal complexes are available [21, 22]. In this article, attention will be focused on specific mechanistic aspects, mainly relating to the role of the bridge in the electron/energy transfer processes. After a short summary of basic con-

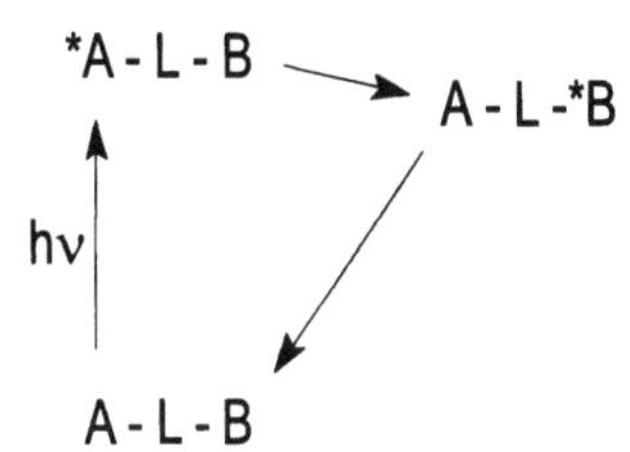

Fig. 2 Schematic representation of electronic energy transfer in a donor–bridge–acceptor system

cepts on electron (Sect. 2) and energy (Sect. 3) transfer, some specific aspects of metal polypyridine complexes as molecular components will be recalled (Sect. 4). In Sect. 5, a number of experimental examples, largely taken from the authors' work, will be discussed.

2 Electron Transfer Mechanisms and Bridge Effects

Electron transfer processes can be described in terms of classical [24–26] or quantum mechanical [27–31] models. From a quantum mechanical viewpoint, both photoinduced electron transfer and charge recombination can be viewed as examples of radiationless transitions between different, weakly interacting electronic states of the supermolecule comprising the donor and the acceptor. The probability of such processes is given by a golden rule expression of the type [27–30]

$$k_{\mathrm{el}} = \frac{4\pi}{h} H_{\mathrm{if}}^{\mathrm{el}\,2} \mathrm{FCWD}^{\mathrm{el}} \tag{1}$$

where $H_{\mathrm{if}}^{\mathrm{el}}$ is the electronic coupling between the two states interconverted by the electron transfer process and $\mathrm{FCWD}^{\mathrm{el}}$ is a thermally averaged vibrational Franck–Condon factor (Franck–Condon weighted density of states). The $\mathrm{FCWD}^{\mathrm{el}}$ term accounts for the combined effects of the nuclear reorganization and driving force and predicts different types of behavior depending on whether the reaction lies in the normal, activationless, or inverted regimes of the classical Marcus theory. Of greater interest in the present context is the role of the electronic factor $H_{\mathrm{if}}^{\mathrm{el}}$.

For a donor–acceptor pair in the absence of any intervening medium (through-space mechanism), the exponential fall-off of the orbital tails leads to an expected exponential decrease of $H_{\mathrm{if}}^{\mathrm{el}}$ with distance

$$H_{\mathrm{if}}^{\mathrm{el}} = H_{\mathrm{if}}^{\mathrm{el}}(0) \exp\left[-\frac{\beta}{2}\left(r_{\mathrm{AB}} - r_0\right)\right] \tag{2}$$

and, neglecting any distance dependence of the FCWD term, to a predicted exponential decay of rate constants with distance.[2]

$$k_{\mathrm{el}} = k_{\mathrm{el}}(0) \exp\left[-\beta\left(r_{\mathrm{AB}} - r_0\right)\right] \tag{3}$$

where r_{AB} is the donor–acceptor distance, $H_{\mathrm{if}}^{\mathrm{el}}(0)$ is the interaction at contact distance r_0, and β is an appropriate attenuation parameter. For donor–acceptor sites in vacuum, estimated values of β are in the range 2–5 $\mathrm{Å}^{-1}$ [31].

---

[2] For real reactions, the FCWD term may exhibit some distance dependence through (a) the reorganizational energy and (b) the driving force. Ideally, a correction for these effects should be done before applying Eq. 3 to experimental results. This point is further discussed in Sect. 5.1.1.

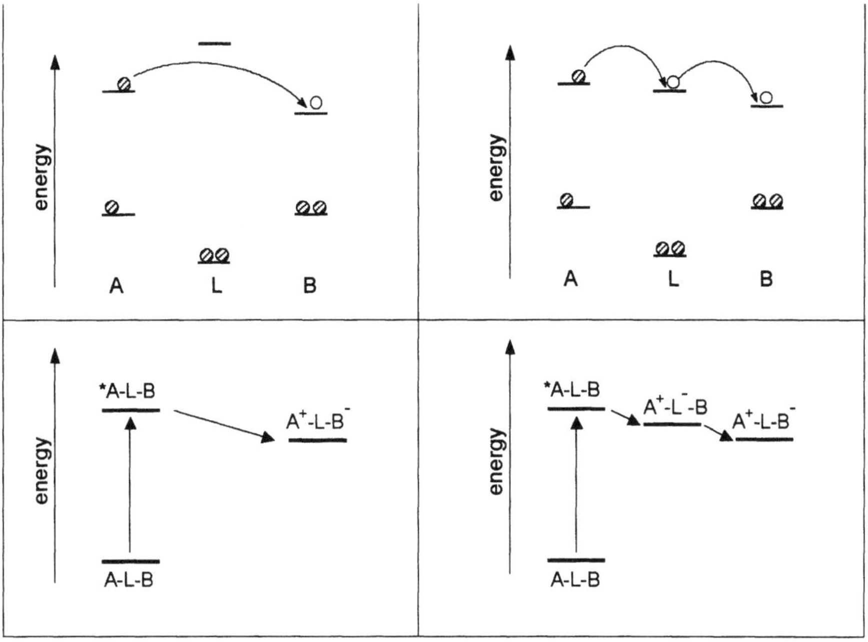

Fig. 3 Schematic representation in terms of orbital (*upper*) and state (*lower*) energy diagrams of superexchange (*left*) and hopping (*right*) mechanisms of photoinduced electron transfer in donor–bridge–acceptor dyads

Thus, at the donor–acceptor distances usually found in dyads, through-space coupling is expected to be negligible.

For photoinduced electron transfer in a donor–bridge–acceptor dyad (Fig. 1), the role of the molecular bridge must be taken into consideration. From this viewpoint, two limiting cases can be identified depending on the relative energies of the donor, bridge, and acceptor levels [20–22, 32]. In a situation such as that depicted in Fig. 3, left, with the empty orbitals of the bridge much higher in energy than those of the donor, the electron is transferred in a single, coherent step from the donor to the acceptor. In this situation, the bridge plays a predominantly (but not exclusively, see later) structural role. In terms of mechanism, this is often described as the *superexchange* regime. If, on the other hand, the LUMO level of the bridge is energetically accessible from the donor orbital (Fig. 3, right), the electron may be first injected into the bridge and then transferred to the acceptor (Fig. 3, right). This situation, where the bridge acts as a real intermediate station for the transferred electron is often indicated as charge *hopping*.

As outlined in the next section, these two mechanisms are expected to yield very different types of experimental behavior, especially in terms of distance dependence of electron transfer probabilities.

2.1 Superexchange

In this mechanism, the electron tunnels directly, in a coherent process, from donor to acceptor. This notwithstanding, the bridge does not act simply as a passive spacer, but rather as an active connector that modulates the magnitude of the donor–acceptor electronic coupling (*through-bond* coupling) [31, 33]. Such through-bond effects are conveniently accounted for in terms of superexchange [20–22, 31, 32, 34–39]. Schematically, the electronic coupling between the initial and final states of the electron transfer process (negligible in the absence of bridge) takes place by mixing with high-energy states of charge transfer character involving the bridge (often called virtual states). Although superexchange pathways involving both donor-to-bridge (electron transfer) and bridge-to-acceptor (hole-transfer) virtual states should be generally considered, one of the two pathways usually predominates. The electron transfer pathway is only considered in Fig. 4. The second-order perturbation expression describing the superexchange coupling is

$$H_{\mathrm{if}}^{\mathrm{el}} = \frac{H_{\mathrm{ie}} H_{\mathrm{fe}}}{\Delta E_{\mathrm{e}}} \tag{4}$$

where H_{ie}, H_{fe}, are the appropriate donor-bridge and bridge-acceptor coupling elements (Fig. 4) and ΔE_{e} is the energy difference between the virtual state and the initial and final state (these energy differences are taken at the transition-state nuclear geometry, where the initial and final state have the same energy). In this expression, the key parameter determining the electron mediating properties of a bridge is the energy denominator: (i) easily reducible bridges, i.e., bridges with relatively low-energy LUMOs, are good electron-transfer superexchange mediators (a similar argument would hold for easily oxidizable bridges if a hole-transfer pathway were considered).

The above picture involving a single virtual state is strictly appropriate for few very simple bridges (e.g., ambidentate ligands such as cyanide,

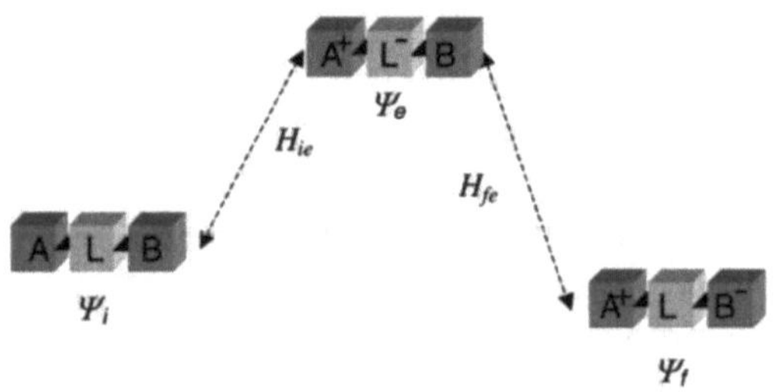

Fig. 4 State diagram illustrating superexchange interaction between a donor (A) and an acceptor (B) through a simple bridging group (L). Ψ_{e} is the electron-transfer "virtual" state involving the bridge. For the other symbols, see text

pyrazine) [36, 38]. In dyads, however, bridges have often a modular structure, being made of a sequence of individual weakly interacting units (see Sect. 5 for several examples). In such a case, not only the chemical nature, but also the length of the bridge will be relevant to its conducting properties. This can be seen by extending the superexchange model to involve virtual charge transfer states localized on each single modular unit, as shown schematically in Fig. 5, where a three-module bridge is considered and, again, for the sake of simplicity, only the electron-transfer pathway is shown. In such a case, the appropriate perturbation expression is

$$H_{\mathrm{if}}^{\mathrm{el}} = \frac{H_{i1}}{\Delta E_{i1}} \frac{H_{12}}{\Delta E_{i2}} \frac{H_{23}}{\Delta E_{i3}} H_{3f} \tag{5}$$

For a bridge involving n identical modular units, the donor–acceptor superexchange coupling takes the form of Eq. 6.

$$H_{\mathrm{if}}^{\mathrm{el}} = \frac{H_{i1} H_{nf}}{\Delta E} \left(\frac{H_{12}}{\Delta E} \right)^{n-1} \tag{6}$$

It is seen that for through-bond superexchange interaction, an exponential dependence on number of modular units in the bridge is again obtained. For linear bridges, this translates into an exponential dependence on donor–acceptor distance. Thus, Eqs. 2 and 3 can still be used, in a phenomenological sense, to describe the distance dependence of donor–acceptor coupling through a modular bridge. The meaning of the various terms of Eq. 2 is seen by comparison with Eq. 6. In this case, r_0 and $H_{\mathrm{if}}^{\mathrm{el}}(0) = (H_{i1}H_{nf})/\Delta E$ represent the donor acceptor distance and the effective coupling for a hypothetical single-module bridge. The attenuation factor $\beta = 2\ln(H_{12}/\Delta E)$ is a bridge-specific parameter depending on (i) the magnitude of the coupling between adjacent modules and (ii) the energy of the electron- (or hole-) transfer states localized on each module. The predicted exponential decay of the electronic

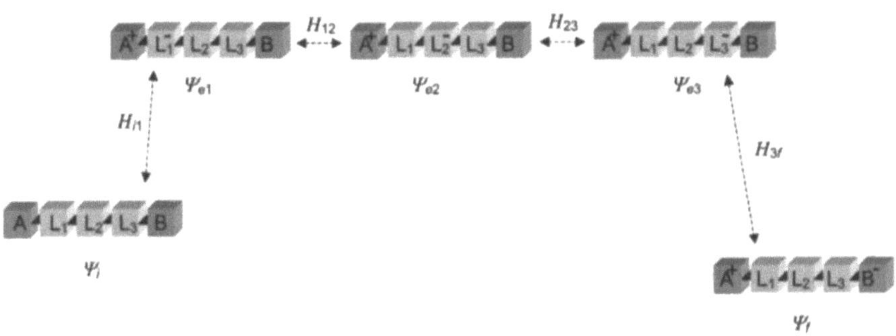

Fig. 5 State diagram illustrating superexchange interaction between a donor (A) and an acceptor (B) through a modular bridge. Ψ_{e1}, Ψ_{e2} and Ψ_{e3} are local electron-transfer "virtual" states

coupling and the effects of (i) and (ii) have been verified in several homogeneous series of dyads containing modular organic bridges of variable length [20].

In conclusion, in the superexchange regime, the bridge plays the role of an active component providing a pathway for donor–acceptor electronic coupling. For this reason, in a very loose sense, bridges are often considered to act as molecular wires. It should be kept in mind that, however, their properties are drastically different from those of a conventional ohmic conductor, particularly in terms of distance dependence of electron transfer probability.

2.2 Charge Hopping

A role of the bridge more similar to that of a real molecular wire is approached in the hopping regime. Here, the energy levels of the bridge are such (Fig. 3, right) that sequential donor-to-bridge and bridge-to-acceptor electron transfer takes place. In the interesting case of a modular bridge (Fig. 6), charge transfer can be split into: (i) charge injection from donor into the bridge, k_{in}; (ii) hopping among (ideally) isoenergetic charge transfer states of adjacent bridge subunits, k_{hop}; (iii) trapping of the electron at the acceptor, k_{t}. The charge hopping along the bridge, which in most cases is likely to be the rate-determining step, can be treated as a diffusive process, characterized in the simplest approximation by an overall rate constant [40]

$$k = k_{\mathrm{hop}} \left(\frac{r_{\mathrm{AB}}}{r_{\mathrm{L}}} \right)^{-2} = k_{\mathrm{hop}} N^{-2} \qquad (7)$$

where r_{AB} is the donor–acceptor distance, r_{L} is the module length, and N is the number of modular units in the bridge. It can be seen that in the hopping regime a relatively weak dependence of charge transfer rates on distance is obtained. The behavior of the bridge is not yet that of a truly homic conductor (where the distance dependence is r_{AB}^{-1}). In this regime, however, given the direct involvement in the charge transfer mechanism and the effectiveness over long distances, the bridge can be considered with more likeliness a molecular wire.

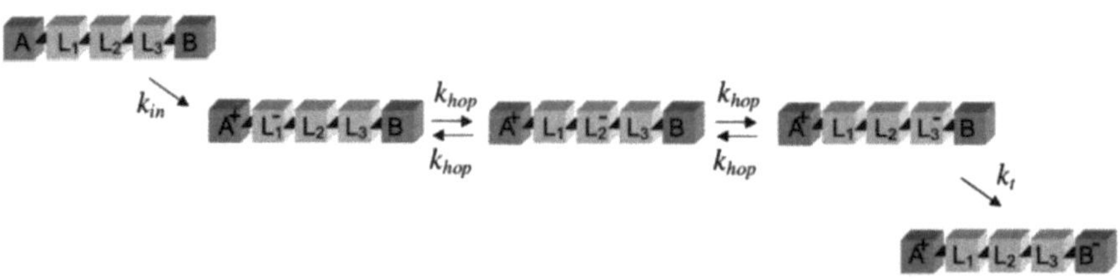

Fig. 6 Charge hopping in a donor–bridge–acceptor system involving a modular bridge

Experimentally, in donor–bridge–acceptor dyads the hopping regime is less common, as it requires potent electron donors and highly conjugated bridges. In cases where these conditions hold, shallow distance dependencies (very low β values when the data are plotted in an exponential law such as Eq. 3) are obtained [41]. A notable case of hopping regime, not related to donor–bridge–acceptor dyads, is the efficient hole transfer taking place in DNA [42].

2.3 Interplay of Mechanisms

In the previous sections, superexchange and hopping mechanisms through modular bridges have been discussed in simplified and rather idealized terms. A caveat is required when transposing these ideas to real systems.

The above discussion on the superexchange mechanism in a modular system, for instance, assumed bridge level energies independent on the number of modules (Fig. 5). This would only be true for ideally non-interacting modules, i.e., a situation where superexchange-mediated electron transfer would not take place ($H_{12} = 0$ in Eq. 6). In any practical system, $H_{12} \neq 0$, and an increase in length is expected to cause a more or less pronounced lowering in the bridge energy levels. As a consequence, the electronic coupling between the bridge and the donor/acceptor sites may change (ΔE not constant in Eq. 6) thus quantitatively affecting (attenuating) the exponential distance dependence of the transfer rate. More important, an increase in bridge length could even cause a switch in mechanism from superexchange to hopping. This is expected to occur when the lowering of the bridge levels brings them below that of the donor. In those cases, an abrupt change is expected to occur at a given bridge length, both in terms of absolute rate values (sudden increase) and distance dependence (from strong to weak). Examples of this behavior have been reported [41, 43].

3 Electronic Energy Transfer

In a supramolecular system, electronic energy transfer (Fig. 2) can be viewed as a radiationless transition between two local electronically excited states of the system.

$$^*A - L - B \rightarrow A - L - {}^*B \tag{8}$$

The rate constant for the energy transfer process is thus given by a golden rule expression similar to that seen above (Eq. 1) for electron transfer.

$$k_{en} = \frac{4\pi}{h}(H_{if}^{en})^2 \mathrm{FCWD}^{en} \tag{9}$$

In Eq. 9, H_{if}^{en} is the electronic coupling between the two excited states interconverted by the energy transfer process and $FCWD^{en}$ is an appropriate Franck–Condon factor.

As for electron transfer, the Franck–Condon factor can be cast either in quantum mechanical [44–46] or in classical [47] terms. In quantum mechanical terms, this factor is a thermally-averaged sum of vibrational overlap integrals, representing the distribution of the transition probability over several isoenergetic virtual transitions (from *A to A, and from B to *B) in the two molecular components. Classically, it accounts for the combined effects of energy gradient and nuclear reorganization on the rate constant. Experimental information on this term can be obtained from the overlap integral between the emission spectrum of the donor and the absorption spectrum of the acceptor.

The electronic factor H_{if}^{en} is a two-electron matrix element involving the HOMOs and LUMOs of the energy-donor and energy-acceptor centers. Following standard arguments [48], this factor can be split into additive *coulombic* and *exchange* terms. The two terms depend differently on various parameters of the system (spin of ground and excited states, donor–acceptor distance, etc.). This leads to the identification of two main energy transfer mechanisms, as pictorially represented in Fig. 7.

The *coulombic* (also called resonance, dipole–dipole, or Förster-type) mechanism [49] is a long-range mechanism that does not require physical contact between donor and acceptor. The most important term within the coulombic interaction is the dipole–dipole term, that obeys the same selection rules as the corresponding electric dipole transitions of the two partners (Fig. 7). Therefore, coulombic energy transfer is expected to be efficient in systems in which the radiative transitions connecting the ground and the

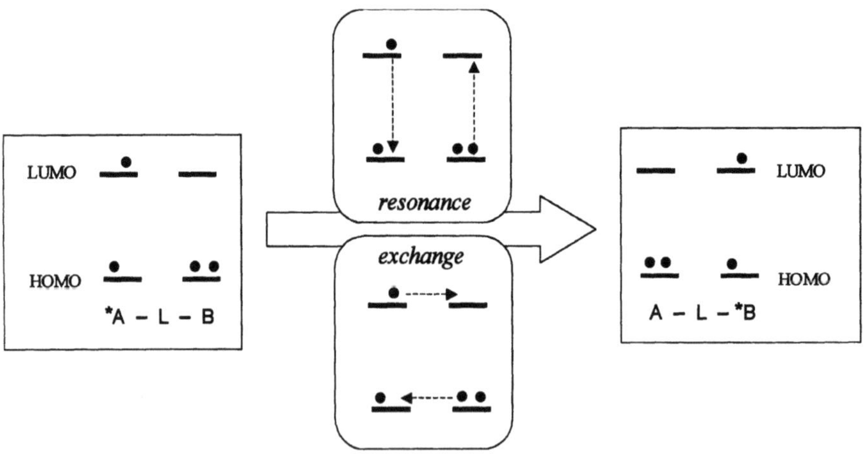

Fig. 7 Pictorial representation of resonance and exchange energy transfer mechanisms

excited state of each partner have high oscillator strength. The relationship between the rate constant for coulombic energy transfer and the spectroscopic and photophysical properties of the two molecular components is given by the classical Förster formula

$$k_{en}^{coul} = 1.25 \times 10^{17} \frac{\Phi_A}{n^4 \tau_A r_{AB}^6} \int_0^\infty F_A(\bar{\nu}) \varepsilon_B \frac{d\bar{\nu}}{\bar{\nu}^4} \tag{10}$$

where Φ_A is the quantum yield of donor emission, n is the solvent refractive index, τ_A is the lifetime of the donor emission, r_{AB} is the distance (in nm) between donor and acceptor, F_A is the emission spectrum of the donor (in wavenumbers and normalized to unity), and ε_B is the decadic molar extinction coefficient of the acceptor. With a good spectral overlap integral and appropriate photophysical parameters, the $(1/r_{AB}^6)$ distance dependence allows energy transfer to occur efficiently over distances (e.g., 50 Å) largely exceeding the molecular diameters. Coulombic energy transfer is therefore also called long-range energy transfer. The typical example of efficient coulombic mechanism is that of singlet–singlet energy transfer

$$^*\text{A}(S_1) - \text{L} - \text{B}(S_0) \rightarrow \text{A}(S_0) - \text{L} - {}^*\text{B}(S_1) \tag{11}$$

In metal complexes, the only excited state of appreciable lifetime is generally the lowest, spin-forbidden excited state [50], so that coulombic energy transfer is only expected to take place in the presence of very strong spin-orbit coupling (e.g., when 5*d* metal complexes are involved).

The exchange (also called Dexter-type) mechanism [49] is a short-range mechanism that requires orbital overlap, and therefore physical contact, between donor and acceptor. The exchange interaction can be visualized as the simultaneous exchange of two electrons between the donor and the acceptor (Fig. 7). The spin selection rules for this type of mechanism reflect the need to obey spin conservation in the reacting pair as a whole. This allows the exchange mechanism to be operative in many cases in which the excited states involved are spin-forbidden in the spectroscopic sense. Thus, the typical example of efficient exchange mechanism is that of triplet–triplet energy transfer

$$^*\text{A}(T_1) - \text{L} - \text{B}(S_0) \rightarrow \text{A}(S_0) - \text{L} - {}^*\text{B}(T_1) \tag{12}$$

Exchange energy transfer from the lowest spin forbidden excited state is expected to be the rule for metal complexes [51].

For reasons similar to those discussed above for electron transfer processes, the rate constant of exchange energy transfer is expected to be sensitive to the nature of the bridge. Using arguments similar to those of superexchange, mixing with electronically excited states localized on the bridge is expected to effectively mediate the donor–acceptor exchange interaction. Again, bridges with low-energy excited states will be particularly efficient.

And again, if the electronically excited states localized on the bridge become lower in energy than that of the donor, a switch from direct one-step energy transfer to stepwise donor-to-bridge and bridge-to-acceptor energy migration could take place. With long, modular bridges, the exchange interaction is expected to fall off exponentially with through-bond distance, or with number of intervening units in the bridge [52–54]. Interestingly, in a homogeneous series of dyads where triplet energy transfer (Eq. 12), electron transfer (Eq. 13), and hole transfer (Eq. 14) processes

$$A^- - L - B \rightarrow A - L - B^- \qquad (13)$$

$$A^+ - L - B \rightarrow A - L - B^+ \qquad (14)$$

could be studied as a function of distance across the same saturated organic bridges, the value of the attenuation factor β (see Eq. 2) for energy transfer was found to be the sum of those for electron and hole transfer [55]. This supports the simple picture of exchange energy transfer as a simultaneous double electron transfer, with an electronic matrix element being proportional to the product of those for the corresponding electron and hole transfer processes.

4
Peculiarities of MLCT Chromophores

In the schematic formulation of Sect. 2, the supramolecular system is represented as an assembly of individual molecular components (e.g., donor, bridge, acceptor). The dissection of a supramolecular system into molecular components is obvious for classical systems where individual molecular species are held together by weak interactions (electrostatic, hydrogen bonding, or host-guest) [56–58]. This situation becomes conceptually more complicated for covalently linked supramolecular systems [59, 60].

Consider, e.g., the two binuclear complexes in Fig. 8. On the basis of simple bonding arguments, this species could be considered as a large molecules. They can usefully be viewed as a *supramolecular species*, however, to the extent to which some of their fragments (*molecular components*) possess intrinsic (i.e., context-independent, or at best slightly perturbed) properties. Properties relevant to photoinduced electron transfer are the redox properties and, for the components that undergo excitation, the chromophoric properties. From this standpoint, the system in Fig. 8a includes two easily identifiable molecular components, the ruthenium polypyridine donor and the rhodium polypyridine acceptor, whose intrinsic chromophoric and redox properties closely resemble those of $Ru(Me_2phen)_2(bpy)^{2+}$ and the $Rh(Me_2bpy)_2(bpy)^{3+}$ model compounds. These two active components are separated by a saturated bis-methylene chain, which can be straightforwardly considered as a bridge (spacer). The distinction could seem

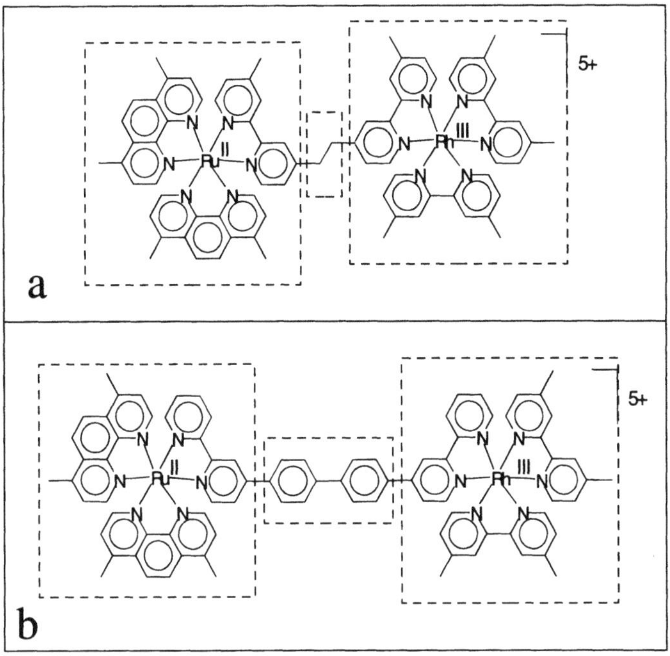

Fig. 8 Examples of covalently linked supramolecular species dissected into molecular components

less obvious for the binuclear complex of Fig. 8b, where the bridging ligand is a chain of interconnected (hetero-) aromatic rings. In practice, however, the chromophoric and redox properties closely resemble again those of $Ru(Me_2phen)_2(bpy)^{2+}$ and the $Rh(Me_2bpy)_2(bpy)^{3+}$ model compounds, indicating that the complex can be reasonably divided into two metal polypyridine molecular components and a bis-phenylene spacer. The reason lies in the steric hindrance that forces adjacent aromatic rings in a twisted conformation, preventing orbital overlap and delocalization along the bridging ligand (this point will be discussed in some detail in Sect. 5.1). Of course, as the delocalization along the bridging ligand increases (e.g., for some of the binuclear species with extended aromatic systems discussed in Sect. 5.2) a supramolecular view becomes more and more critical.

Aside from the above-mentioned general problems, some peculiar aspects of supramolecular systems containing coordination compounds deserve a short discussion. A coordination compound is by definition a complex species formed by combination of metal and ligands. This complex nature has relevant effects on (i) their redox and excited-state properties and (ii) the electron/energy behavior as molecular components of supramolecular systems. Some of these features can be discussed taking the dyad of Fig. 8b as an example.

The Ru(II)-based molecular component of the dyad in Fig. 8b can be adequately modeled by $Ru(bpy)_3{}^{2+}$, the archetypal chromophore of inorganic photochemistry [61, 62]. In $Ru(bpy)_3{}^{2+}$, the HOMO is largely metal t_{2g} in character while the LUMO is largely bpy π^* in character. This predominant localization of the frontier orbitals has several interesting consequences. First, oxidation and reduction of the unit involve different sites: $Ru(bpy)_3{}^{3+}$ is a true Ru(III) complex, whereas the $Ru(bpy)_3{}^{+}$ is not a Ru(I) species, but rather a Ru(II) complex with two neutral ligands and one ligand radical anion [63]. Second, the lowest excited states of $Ru(bpy)_3{}^{2+}$ are of metal-to-ligand charge transfer (MLCT) character, with an electron distribution resembling a Ru(III) complex with two neutral ligands and one ligand radical anion. Despite the presence of appreciable spin-orbit coupling, the MLCT states can be considered as having singlet or triplet spin multiplicity. The lowest singlet is responsible for the intense visible absorption while the lowest triplet is responsible for the characteristic phosphorescence emission. The relatively small singlet-triplet energy difference (ca. 0.4 eV) is a consequence of the charge transfer nature of the transition. Within the triplet lifetime, fast equilibration between the three MLCT states localized on different ligands can be usually assumed [64]. The fractional population of such localized states (33% in symmetric complexes such as $Ru(bpy)_3{}^{2+}$) can be substantially unbalanced in less symmetric cases. In ligand-bridged bimetallic systems, a relevant question may be whether the predominant population is on the bridging ligand or on the ancillary ones.[3] This depends on the relative LUMO energies (or redox potentials) of the tree ligands. In the Ru(II) polypyridine unit of the dyad in Fig. 8b, because of the appreciable (albeit small) conjugation with the polyphenylene chain, the lowest energy site for the excited electron is definitely the bridging bipyridine.

For typical (e.g., organic) donor–acceptor systems, simple HOMO-LUMO pictures such as those of Figs. 3 and 7 are often used to describe photoinduced electron or energy transfer. When coordination compounds are used as A and/or B molecular components, such kind of pictures might be overly simplistic given the peculiar localization of HOMOs and LUMOs. Figure 9a schematically shows the redox sites involved in MLCT excitation (process 1), photoinduced electron transfer (process 2) and charge recombination (process 3) for a typical Ru(II)–Rh(III) dyad. These localization aspects are likely to be important in the analysis of intramolecular electron transfer rates, where intercomponent distance and electronic coupling play a crucial role.

Similar arguments apply to the case of intercomponent energy transfer involving MLCT chromophores. In a Ru(II)–Os(II) dyad such as that of Fig. 9b, exchange energy transfer following MLCT absorption by the Ru(II) component (process 1) and leading to MLCT emission from the Os(II) component

[3] The localization of the MLT state on bridging or peripheral ligands may have relevant consequences on energy transfer kinetics. A clear example is discussed in Sect. 5.1.2.

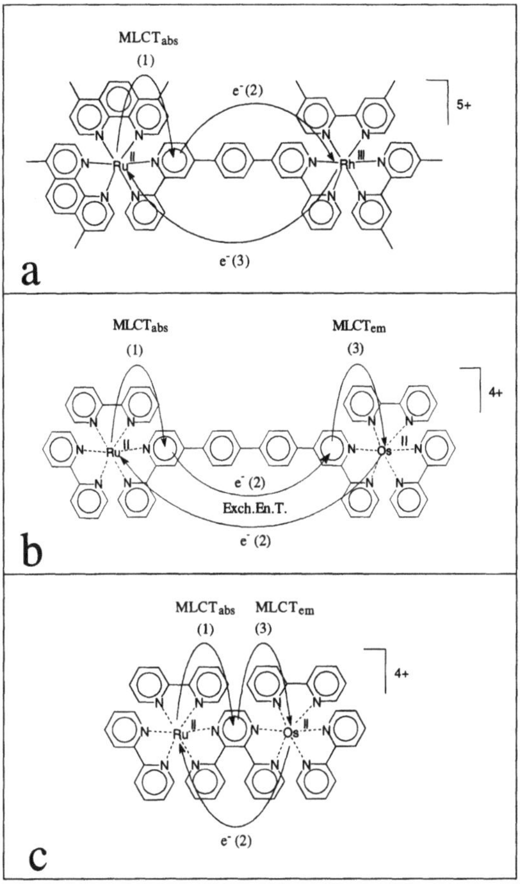

Fig. 9 Schematic representation of the one-electron processes involved in photoinduced electron and energy transfer in bimetallic dyads (for detailed discussion, see text)

(process 3) can be visualized as two simultaneous cross-electron-transfer processes (process 2): between the two bpy ends of the bridging ligand, and between the two metal centers. It should be noticed that, with short bridges such as that of the Ru(II)–Os(II) dyad of Fig. 9c, what is formally (by analogy to the previous case) an intercomponent energy transfer phenomenon in fact amounts to a simple metal-to-metal electron transfer (process 2).

5 Case Studies

Studies of intercomponent energy/electron transfer processes have been performed on a great variety of bimetallic dyads [21, 22]. In this review we discuss in particular the behavior of dyads containing two specific classes of bridges: polyphenylenes and poly-quinoxalines. Such bridges are longitudinally rigid, so as to guarantee good structural definition of the dyad. They are conjugated systems favoring, in principle, energy/electron transfer over long distances. As shown below, however, the behavior of the two types of bridge is quite different. The polyphenylene systems behave as ideal superexchange mediators, while in the poly-quinoxaline series injection behavior and other mechanistic complications are encountered.

5.1 Dyads with Poly-*p*-phenylene Bridges

Dyads with poly-*p*-phenylene bridges have attracted much interest because in theses systems the bridge is made of a sequence of individual weakly coupled rigid units. This allows one to obtain families of dinuclear compounds in which the metal–metal distance can be tuned without changing the nature of active units and to perform systematic studies on the distance dependence of electron and/or energy transfer rates. Because of synthetic difficulties related to solubility problems, in most of the systems studied the number of phenylene spacers was limited to one or two units [21, 22, 65–73]. Recently, however, more extended systems have been reported [21, 74, 75].

5.1.1 Ru(II)–Rh(III) Systems

A series of Ru(II)–Rh(III) dyads bridged by rod-like polyphenylene spacers of various length have been synthesized and studied in the last few years [21, 68, 69, 76]. In these systems, the two metal units are suitable pairs for oxidative photoinduced electron transfer, the Ru(II) polypyridine unit playing the role of photoexcitable donor (A) and the Rh(III) polypyridine unit the role of electron acceptor (B in Fig. 1). The photoexcitable donor fulfills the following requisites: (i) low-energy intense absorption (i.e., be a visible chromophore), (ii) strong excited-state reducing character, (iii) relatively long excited state lifetime (to support relatively slow electron transfer processes) and (iv) easily detectable emission (for facile excited-state monitoring). Taking $Ru(Me_2phen)_2(Me_2bpy)^{2+}$ (**1**) as a paradigmatic case, the relevant properties are: $E^{0-0} = 2.16$ eV, $E^0(A^+/{}^*A) = 1.02$ V, $E^0(A^+/A) = 1.14$ V vs. SCE, $\lambda_{max}^{em} = 610$ nm, $\Phi = 0.11$, $\tau = 1.8$ µs [73].

The electron acceptor satisfies specific requirements that it: (i) be practically non-chromophoric (so as to avoid absorption of excitation light) and (ii) be a good ground-state electron acceptor $[E^0(B/B^-) > E^0(A^+/{}^*A)]$, (iii) lack of low-energy excited states $[E^{0-0}({}^*B) < E^{0-0}({}^*A)$, in order to avoid energy transfer quenching], (iv) be kinetically stable in the reduced form (at least in the time scale of the charge recombination process). For $Rh(Me_2bpy)_3{}^{3+}$

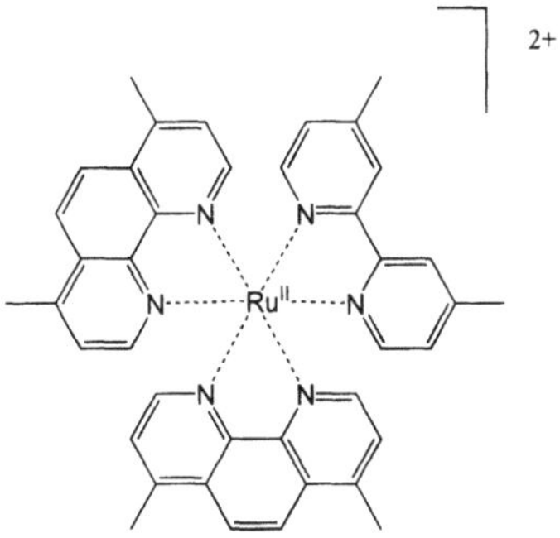

(1)

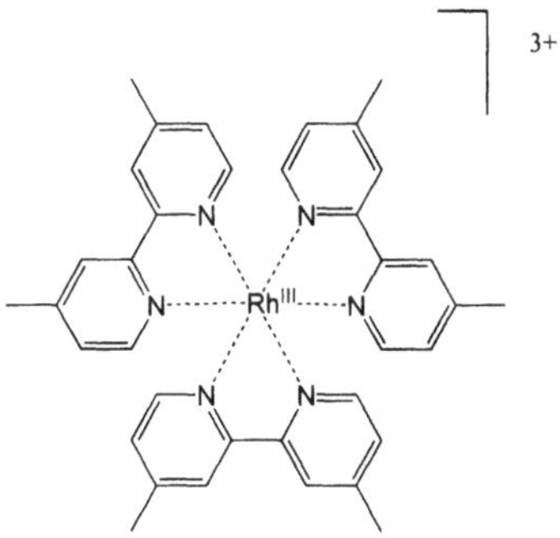

(2)

Formula (**2**) as a paradigmatic case, relevant properties are: $E^{0-0} = 2.77$ eV, $E^0(\mathrm{B/B^-}) = -0.89$ V vs. SCE, $\tau = 3$ ns [76].

The series of dyads studied is shown in Scheme 1. In this series the chromophoric and acceptor units were of the tris-bipyridine (or phenanthroline) type [21] or of the bis-terpyridine type [68, 69]. The general properties of Ru(II)–Rh(III) dyads can be discussed taking dyad (**5**), with one phenyl as spacer, as an example.

The absorption spectrum of the dyad, is practically the superposition of the spectra of the isolated mononuclear model complexes (**1**) and (**2**), as expected for weak intercomponent interaction. The Rh(III) complex does not absorb in the visible region, where the spectrum of the dyad is characterized by the typical metal-to-ligand charge transfer (MLCT) transitions of the Ru(II) component. Therefore selective (100%) excitation of the Ru(II) chromophore can be easily performed in the visible, whereas partial excitation of the Rh(III) component can be achieved in the ultraviolet (where overlapping ligand-centered, LC, bands of both molecular components are present). The energy level diagram for this dyad is shown in Fig. 10. The diagram holds for the whole series of dyads studied, with minor quantitative differences

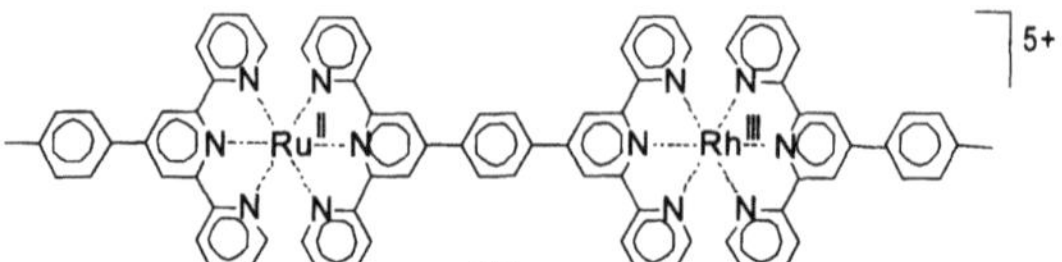

(3)

$k = 3.0\text{-}4.2 \times 10^9\ s^{-1}$

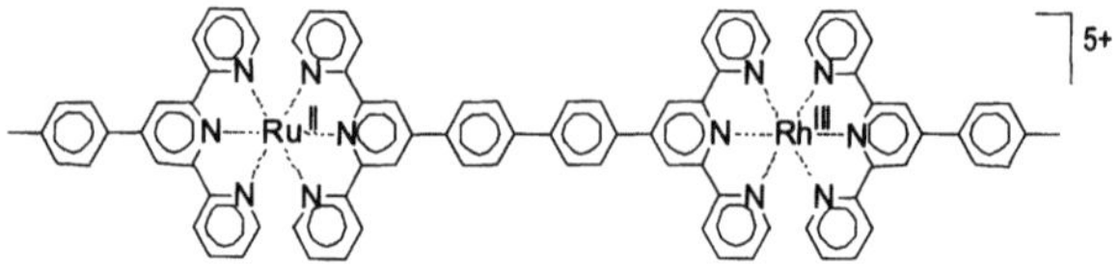

(4)

$k\quad 3.0 \times 10^9\ s^{-1}$

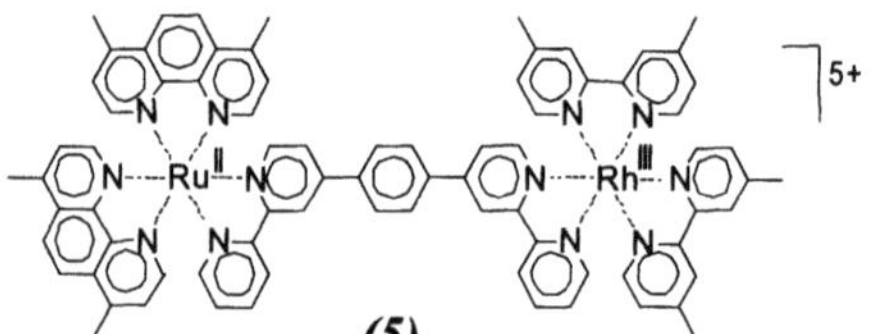

(5)

$k = 3.0 \times 10^9\ s^{-1}$

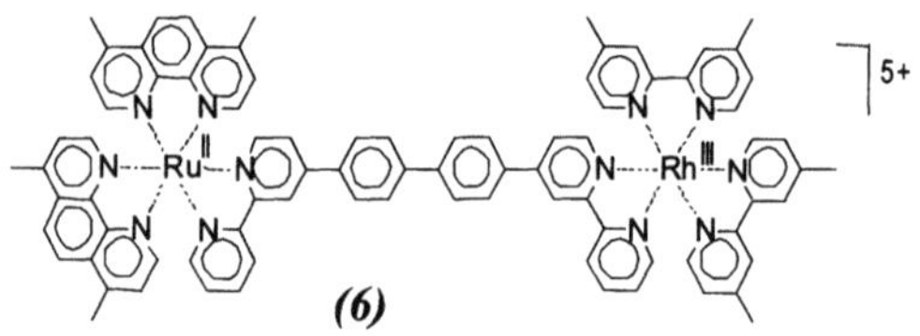

(6)

$k = 4.3 \times 10^8\ s^{-1}$

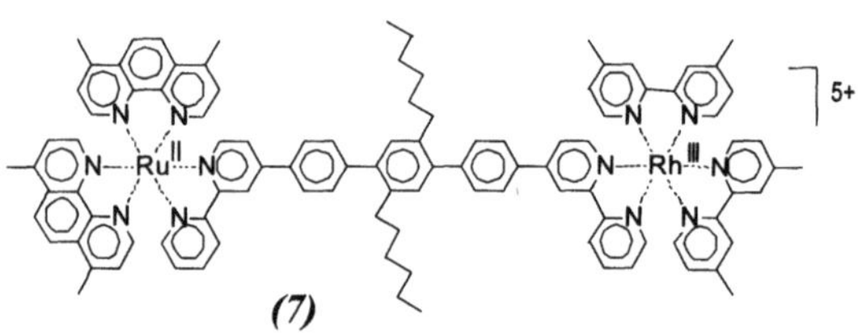

(7)

$k = 1.1 \times 10^6\ s^{-1}$

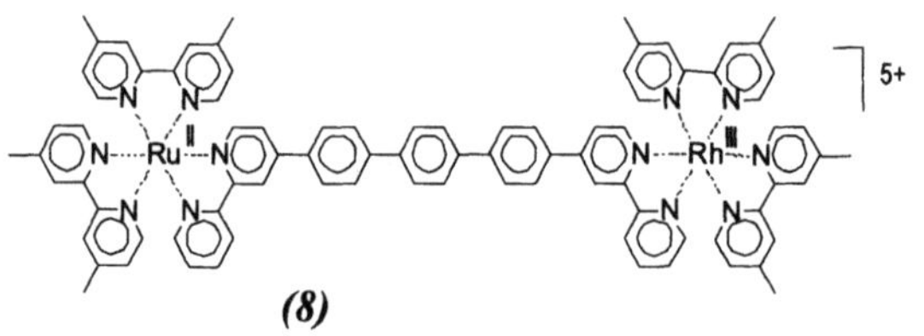

(8)

$k = 1.0 \times 10^7\ s^{-1}$

Scheme 1

between the bpy (or phen) complexes and the tpy ones. Indicated are the common photophysical processes taking place within each molecular components, i.e., prompt intersystem crossing to form the long-lived triplet states.

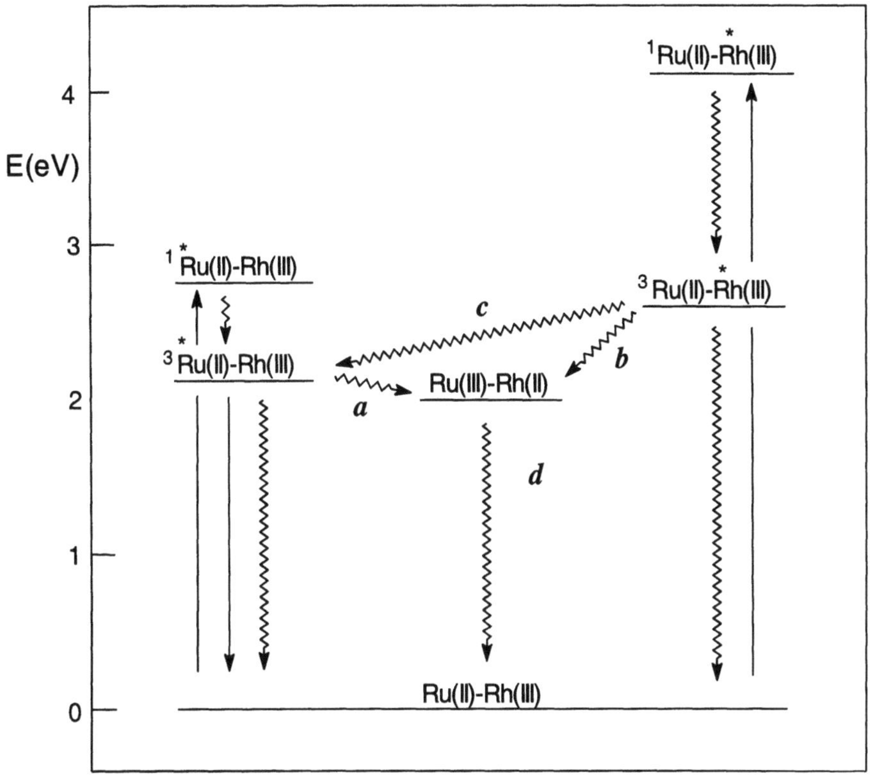

Fig. 10 Energy level diagram of the Ru(II)–Rh(III) dyad (**5**)

Also indicated are a number of thermodynamically allowed intercomponent processes, including:

(a) *Ru(II) – Rh(III) → Ru(III) – Rh(II)
oxidative quenching of the excited Ru(II)

(b) Ru(II) – *Rh(III) → Ru(III) – Rh(II)
reductive quenching of the excited Rh(III)

(c) Ru(II) – *Rh(III) → Ru(III) – Rh(II)
energy transfer from Rh(III) to Ru(II)

(d) Ru(III) – Rh(II) → Ru(II) – Rh(III)
charge recombination

In principle, all of these processes can be time-resolved under favorable experimental conditions [76]. With this series of poly-*p*-phenylene bridged systems, attention was focussed on the slightly exergonic oxidative quenching following Ru(II) excitation (process a in Fig. 10) and the emphasis was placed on the study of bridge and distance effects. This type of study is usually performed by exciting the chromophoric donor (Ru(II) component) with appropriate light pulses and by observing electron transfer to the acceptor (Rh(III) component) by means of time-resolved spectroscopy. In analyzing the results of these studies, it should be kept in mind that the energetics of photoinduced electron transfer (see, e.g., Fig. 10) is practically the same for the all systems investigated.

The rate constants measured for photoinduced electron transfer process are reported in Scheme 1 and indicate a general decrease of electron transfer rates with increasing bridge length [21]. Of particular interest is the comparison between the omogeneous dyads containing tris-bypyridine (or phenanthroline) species as chromophore ((**5**), (**6**), (**8**)). The rate constants are displayed on a logarithmic plot as a function of the metal–metal distance in Fig. 11. It can be seen that the three systems with unsubstituted polyphenylene spacers exhibit a nice exponential dependence of rates on distance. This is the behavior predicted by Eqs. 1 and 2 on the basis of the exponential decay of the electronic factor H_{if}, if the distance dependence of the FCWD term can be neglected. This type of dependence is consistent with a superexchange mechanism (see Sect. 2.1).

An interesting result is the fact that dyad (**7**), which is identical to (**8**) except for the presence of two solubilizing hexyl groups on the central phenylene ring, undergoes a photoinduced process 10 times slower than its unsubstituted analog. This is related to the notion that in a superexchange mechanism the rate is sensitive to the electronic coupling between adjacent molecules of the spacer (eg. H_{12}, H_{23} in Eq. 5), and that in polyphenylene bridges this coupling is a sensitive function of the twist angle between adjacent spacers [77, 78]. It is worthy to note that the planes of the rings of unsubstituted adjacent phenylene unit form an angle of ca. 20–40° [79, 80] because the steric repulsion between the ortho hydrogens. Substitution of the rings leads to an increase in the twist angle (ca. 70°) [79] and, as a conse-

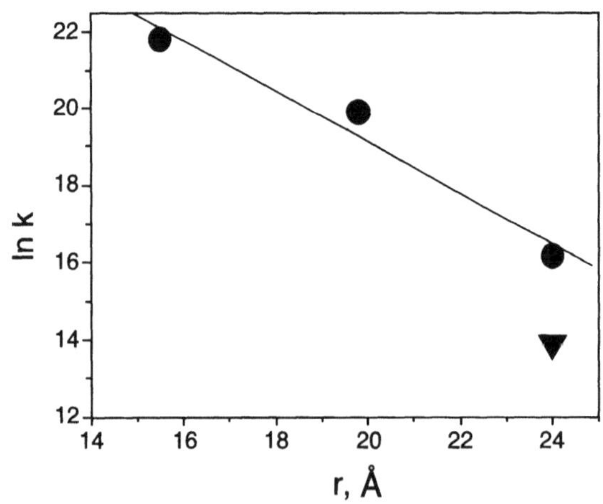

Fig. 11 Distance dependence of photoinduced electron transfer rates in dyads (**5**), (**6**), (**8**) (*circles*) and (**7**) (*triangle*)

quence, a decrease of electronic coupling between adjacent phenyl rings that causes a drastic slowing down of the electron transfer process.

From the slope of the line in Fig. 11 an attenuation factor β (Eq. 3) of 0.65 $Å^{-1}$ can be calculated. It is important to note that this value is obtained neglecting the distance dependence of the FCWD term (Eq. 1). The two possible causes of distance dependence of the FCWD term are the driving force and the solvent reorganizational energy, λ [25, 81]. From this viewpoint, a useful feature of the Ru(II)–Rh(III) dyads is the charge-shift nature of the process (it leads from a 2 + /3 + to a 3 + /2 + species). Thus, at difference with typical organic dyads, the driving force is independent on the donor–acceptor distance. On the other hand λ is expected to increase slightly with distance. Its contribution to the observed rate decrease is presumably relatively small, but is difficult to evaluate exactly as it depends on the assumed partitioning of λ into inner and outer parts. Therefore the β value of 0.65 $Å^{-1}$ should be regarded as an upper limiting value for the attenuation factor of the intercomponent electronic coupling. This value is much lower than those observed for rigid saturated bridges (0.8–1.2 $Å^{-1}$) [20, 33, 82–84], and similar to that obtained for electron transfer across oligophenylene spacers between porphyrins by Helms et al. [85] and more recently by Weiss et al. [43] for organic donor–acceptor systems.

It is instructive to compare the electron transfer rate constant observed for dyad **(5)** ($k = 3.0 \times 10^9\ s^{-1}$) of the Scheme 1 with that measured for dyad **(9)** ($k = 1.7 \times 10^9\ s^{-1}$). The two dyads are similar in driving force and reorganizational energies (FCWD in Eq. 1). Thus the main effective differences, lies in the bridge. It is seen that, although the metal–metal distance is longer in **(5)** (15.5 Å relative to 13.5 Å for **(9)**) the electron transfer is faster across the phenylene spacer than across the methylene one. This result demonstrates that the phenylene group is a better conducting spacer as compared to methylene. This is no doubt related to the lower energy of the LUMO of the phenylene group, facilitating superexchange interaction (Eq. 4).

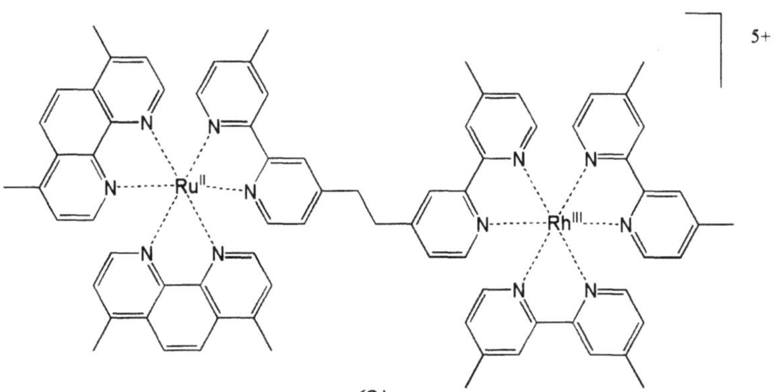

(9)

A last observations is worth of comment. Dyads (**3**) and (**5**), despite the vast difference (three orders of magnitude) in lifetimes of the chromophores involved, give identical electron transfer rates. This is experimentally gratifying and confirms that the rate determining factors are practically the same for dyads of the bpy (or phen) and tpy families.

In conclusion the study of the Ru(II)–Rh(III) dyads in Scheme 1 has shown that rigid polyphenylene bridges are efficient mediators of long-distance photoinduced electron transfer. Coupled with the long lifetime of the excited Ru(II) chromophore, such bridges could, in principle, sustain electron transfer over very long distances (e.g., extrapolating from Fig. 11, ≥ 40 Å). In principle, the dependence of the electronic coupling on the inter-spacer twist angle could be used for tuning or switching purposes.

5.1.2
Ru(II)–Os(II) Systems

Binuclear systems containing ruthenium(II) and osmium(II) complexes as active metal units represent by far the most extensively investigated class of compounds for the study of electronic energy transfer processes. In most of the dyads studied, the Ru-based and Os-based chromophoric components are tris-bipyridine or bis-terpyridine complexes connected by a large variety of bridging ligands [21, 22, 65, 66, 86–89]. For both metal components, the lowest excited states are triplet states of metal-to-ligand charge transfer (MLCT) nature. The Os(II) unit is easier to oxidize than Ru(II) unit, and thus the MLCT levels lie at higher energy in the Ru-based component than in the Os-based component. Therefore the Ru(II) units play the role of donor and the Os(II) units of acceptor in an intercomponent energy transfer process (Eq. 15).

$$^{*}\mathrm{Ru(II)} - \mathrm{L} - \mathrm{Os(II)} \rightarrow \mathrm{Ru(II)} - \mathrm{L} - {}^{*}\mathrm{Os(II)} \qquad (15)$$

Dyads containing bridging ligands with oligophenylene spacers have been studied by various authors [21, 22, 65, 66] but only recently new synthetic strategies based on a modular approach have been developed that allowed to obtain families of dyads with high number of phenylene units [74, 75].

The series of the binuclear complexes (**10**) investigated by Schlicke et al. [74] represent the first example of a systematic study on distance dependence of energy transfer rates for Ru(II)-Os(II) dyads. These systems of nanometric dimension (the metal to metal distance is 4.2 nm for longer spacer) contain a central phenylene unit which bears two alkyl chains for solubility reason and a progressively increasing number of unsubstituted phenylene units.

The absorption spectra and luminescence properties have been investigated in acetonitrile solution at 293 K and in butyronitrile rigid matrix at 77 K. Besides complete quenching of the fluorescence of the oligophenylene

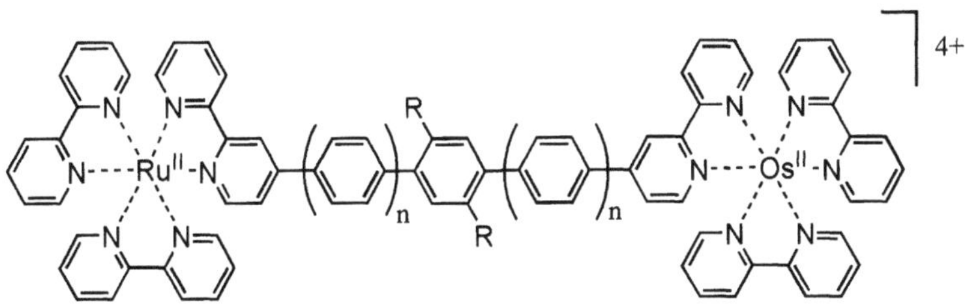

(10)

n = 1, 2, 3

spacers, quenching of the phosphorescence of the Ru(II) based unit and parallel sensitization of the phosphorescence of the Os(II)-based unit were observed, indicating that energy transfer from $Ru(bpy)_3{}^{2+}$ to the $Os(bpy)_3{}^{2+}$ unit takes place (Eq. 15). The rate constant of the process is practically independent of temperature and decreases with increasing length of the oligophenylene spacer. The plot of $\ln k_{en}$ vs. the metal–metal distance is shown in Fig. 12. As one can see, a linear dependence was observed according with expectations based on a Dexter-type mechanism (Sect. 3). The plot yields an attenuation factor (β) of 0.32 $Å^{-1}$. This value quantifies the behavior of polyphenylene spacers as good energy transducers.

A new series of Ru(II)–Os(II) dyads (**11**) have been synthesized very recently by Welter and coworkers [75].

These compounds are very similar to the previous ones except for the fact that the metal centers $Ru(bpy)_3{}^{2+}$ and $Os(bpy)_3{}^{2+}$ are linked by a bridg-

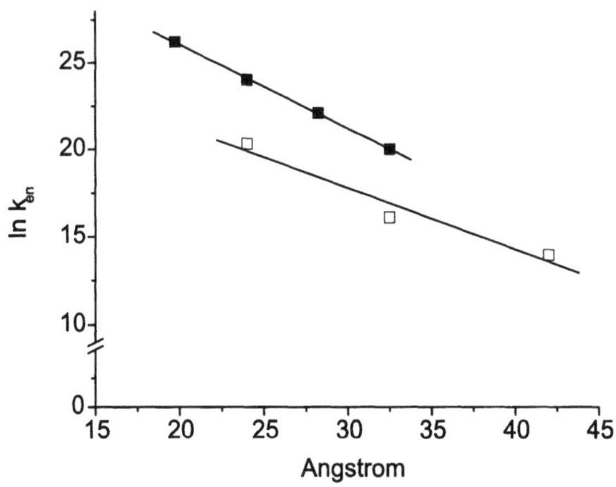

Fig. 12 Plot of $\ln k_{en}$ vs. the metal-metal distance, (■) dyads (**11**) ($n = 2, 3, 4, 5$) with bare phenylene units, (□) dyads (**10**) ($n = 1, 2, 3$) with n-hexyl chains on the central phenylene unit

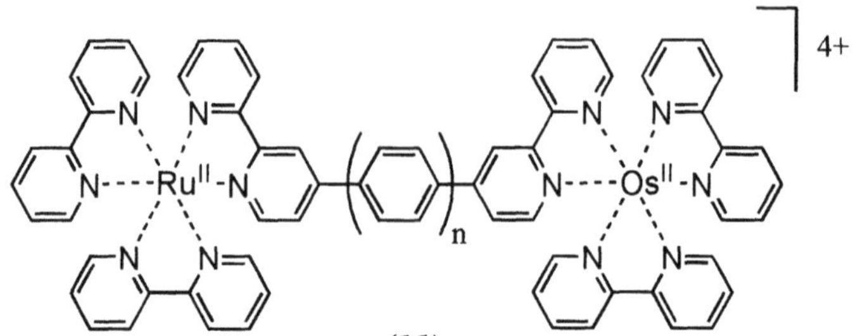

(11)

n = 2, 3, 4, 5

ing ligand where the spacer is made only of unsubstituted phenylene units ($n = 2$–5). The elimination of the bulky solubilizing groups permits examination of the effects of inter-phenylene twist angle on energy transfer rate. The rate constants, measured by time resolved emission and absorption spectroscopy, are reported in Fig. 12 and compared with the data obtained for the dyads **(10)** in which the central phenylene unit is substituted. Consider, e.g., the compounds containing three or five spacers for the two series: the driving force and the metal to metal distance is the same while a difference of rate constant of 40 fold has been observed in both cases. This is a consequence of the increase in twist angle upon substitution that reduces the electronic coupling along the spacer. Interestingly a similar same decrease of rate has been reported for photoinduced electron transfer process in analogous Ru(II)–Rh(III) polyphenylene dyads **(7)** and **(8)** (Sect. 5.1.1). Inspection of Fig. 12 clearly indicates that for both series of dyads the plots of $\ln k_{en}$ vs. metal–metal distance are linear confirming that energy transfer process across polyphenylene spacers takes places with a Dexter-type (exchange) mechanism. As expected on the basis of the structural similarity (in both cases, elongation only involves unsubstituted phenylene spacers), the slope of the lines for the two series of dyads is practically the same (a β value of ca. 0.4 $Å^{-1}$ is obtained in this case).

It is interesting to note that the β values for energy transfer in these Ru(II)–Os(II) complexes (0.34–0.40 $Å^{-1}$) are very similar to that obtained for photoinduced electron transfer in Ru(II)–Rh(III) complexes with the same bridges (0.65 $Å^{-1}$). On general grounds, as discussed in Sect. 3, β values for exchange energy transfer are expected to be larger than those for electron transfer [55]. In fact, this coincidence may be fortuitous as, because of the distance dependence of the reorganizational energy, the real β value for the Ru(II)–Rh(III) system is expected to be lower than the experimental one (see discussion in Sect. 5.1.1). A comparison between the behavior of dyads **(11)** and dyads **(12)** is instructive in order to highlight the importance of the structural and electronic factors on the energy transfer rates.

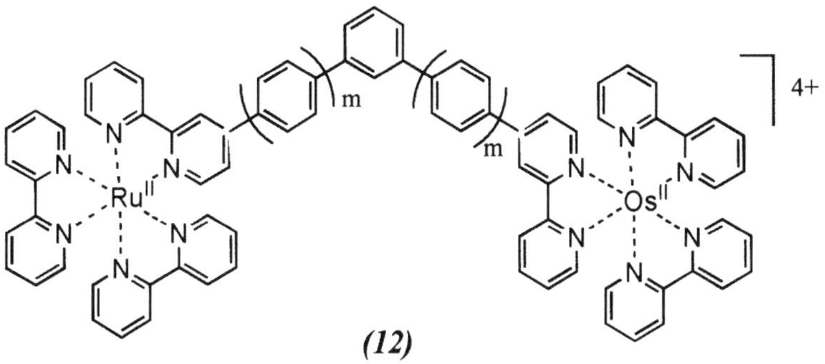

(12)

m = 1, 2

Consider, e.g., the dyads with three and five phenylene units for both series (De Cola L, personal communication): energy transfer is faster across the *para* substituted ($k = 2.77 \times 10^{10}\ s^{-1}$ and $4.90 \times 10^{8}\ s^{-1}$ for $n = 3, 5$ respectively) than across *meta* substituted ones ($k = 1.32 \times 10^{9}\ s^{-1}$ and $6.67 \times 10^{7}\ s^{-1}$ for $n = 3, 5$ respectively). The results clearly demonstrate that energy transfer occurs through-bond (the through-space distance is larger for systems **(11)** than for systems **(12)** and that *para*-phenylene spacers are better energy transducers as compared to *meta*-phenylene ones.

Another interesting comparison is between two series of related Ru(II)–Os(II) polyphenylene dyads containing bis-terdentate complexes as metal based components. The first series of dyads [65, 70, 71] contains bis-terpyridine species **(13)**. In the second series [71, 72] tpy chelating sites of the bridging ligand are substituted by cyclometallating units **(14)**. The results show that the rates of energy transfer are orders of magnitude slower for complexes **(14)** with respect to complexes **(13)** with the same number of phenylene spacers (e.g., with two phenylene spacers, $k < 2 \times 10^{7}\ s^{-1}$ and $k = 5 \times 10^{10}\ s^{-1}$ for cyclometallated and non-cyclometallated system, respectively). The reason for this behavior is the different localization of MLCT states involved in the transfer process: in complexes **(13)** the lowest MLCT states of each chromophore involve the bridging ligand whereas in complexes **(14)** these states are local-

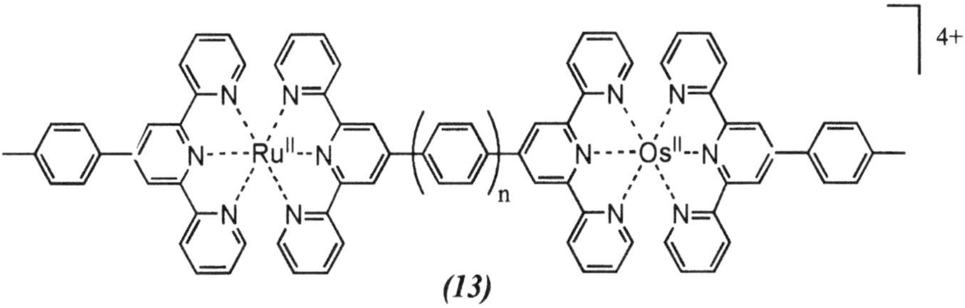

(13)

n = 0, 1, 2

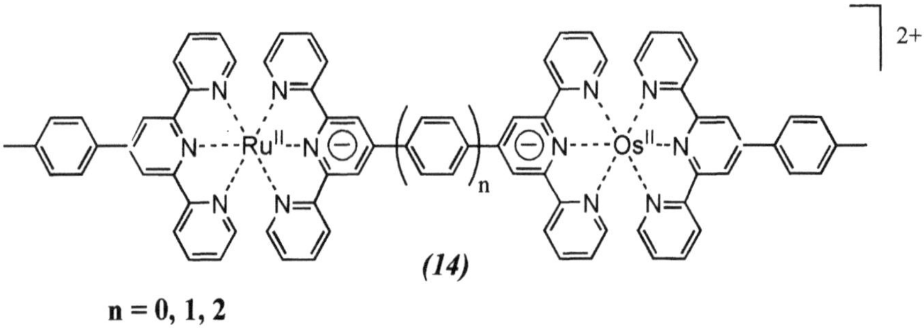

ized on the peripheral ligands. Therefore, in the latter case the energy transfer pathway is longer. This result highlights the crucial role of excited-state localization in determining the energy transfer rate.

5.2 Dyads with Extended Aza-aromatic Bridges

An interesting class of molecular bridges that have been recently used in the construction of binuclear and polynuclear complexes is represented in Fig. 13. Because of the presence of fused pyrazine and phenyl rings as characteristic motif, these bridges can be loosely termed poly-quinoxaline bridges. With their extended aromatic character, these ladder-type ditopic ligands cannot be dissected, as done for most of the previously discussed examples, in bpy-like binding sites (belonging to the metal-based molecular components) and a central spacer part. Rather, the ligand should be treated as a single entity, with fully delocalized molecular orbitals (see, e.g., the LUMO

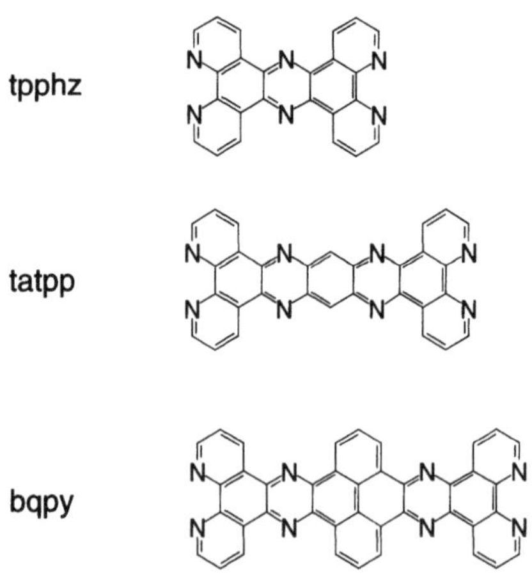

Fig. 13 Representative poly-quinoxaline bridging ligands

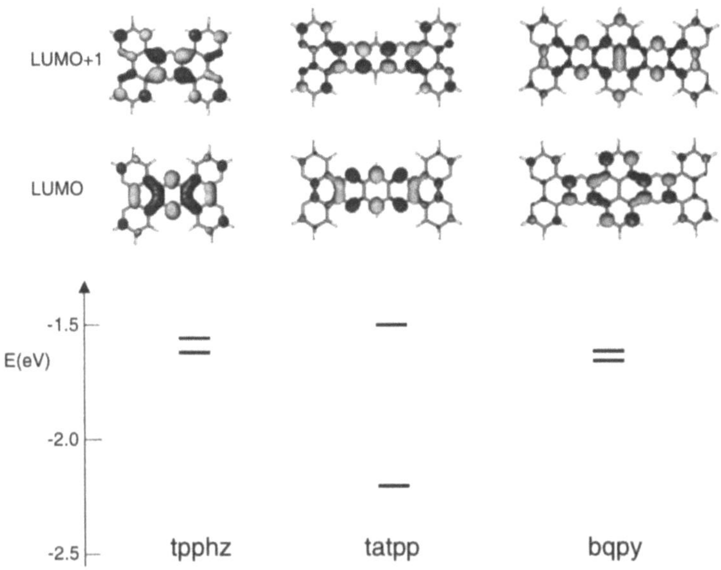

Fig. 14 Low-lying unoccupied molecular orbitals for the tpphz, tatpp, and bqpy ligands. Relative energies as obtained from PM3 calculations

and LUMO + 1 depicted in Fig. 14). This section highlights some of the peculiar problems that are encountered in the analysis of electron/energy transfer processes across this class of bridging ligands.

5.2.1
tpphz Complexes

Binuclear complexes with this bridging ligand have been studied in considerable detail. Information from fast (ns) and ultrafast (fs-ps) studies is available on both the homonuclear Ru(II) and Os(II) species. Moreover, the heteronuclear Ru(II)–Os(II) and Ru(II)–Os(III) species have been investigated, with the aim of unraveling the complex mechanisms of energy and electron transfer processes.

5.2.1.1
$[(bpy)_2Ru(tpphz)Ru(bpy)_2]^{4+}$

Spectroscopic, electrochemical, and photophysical studies with ns time resolution [90–92] yielded the following picture of $[(bpy)_2Ru(tpphz)Ru(bpy)_2]^{4+}$:

1. in the ground state, the binuclear complex can be described as a supramolecular system with relatively weak intercomponent interaction;

2. in the binuclear species, the lowest emitting metal-to-ligand charge transfer (MLCT) excited states involves the tpphz bridge;
3. consistent with the presence of two closely spaced ligand π^* orbitals with different electron density patterns (Fig. 14), two types of MLCT states involving the bridging ligand are present in this complex: $MLCT_1$, with no electron density on the central nitrogens and substantial density on the coordinating atoms, and $MLCT_0$ with substantial electron density on central nitrogens and little density on the coordinating atoms. For simplicity, these MLCT states can be referred as phenanthroline-like and pyrazine-like and can be schematically depicted as in Fig. 15;[4]
4. states of type $MLCT_1$ are involved in absorption;
5. at room temperature, emission takes place from states of $MLCT_0$ type; the $MLCT_0$ states are substantially stabilized in polar solvents, leading to pronounced red shift and lifetime shortening of the emission.

More recently, ps [94] and fs [95] time-resolved experiments have permitted the direct observation of the $MLCT_1 \rightarrow MLCT_0$ interconversion. The formation of the $MLCT_0$ state is clearly detected as an increase on opti-

[4] A similar situation holds for ruthenium complexes of the related dppz ligand [93].

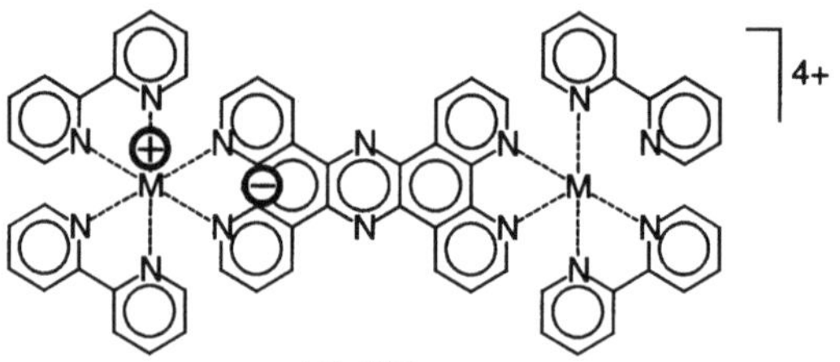

$MLCT_1$

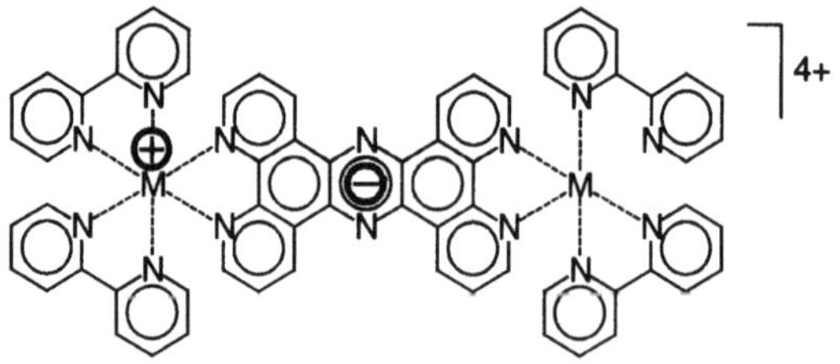

$MLCT_0$

Fig. 15 Schematic representation of the two types of metal-to-bridge charge transfer excited states in tpphz binuclear complexes: $MLCT_1$, with main electron density on the phenathroline-like moiety of the bridge; $MLCT_0$, with main electron density on the pyrazine-like moiety of the bridge

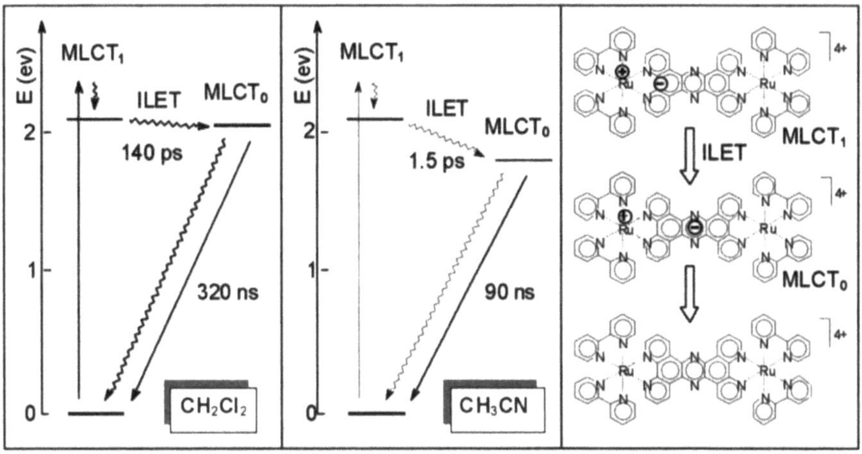

Fig. 16 Photophysical mechanism for $[(bpy)_2Ru(tpphz)Ru(bpy)_2]^{4+}$ in CH_2Cl_2 (*left*) and CH_3CN (*center*). Schematic picture of the changes in charge distribution along the photophysical pathway (*right*)

cal density with maximum at 610 nm. The kinetics of formation and decay of this state are solvent-sensitive, as a consequence of the above-mentioned (point 5) energy shifts. The photophysical behavior can be summarized as shown schematically in Fig. 16 [95].

In Fig. 16, the $MLCT_1 \rightarrow MLCT_0$ interconversion is indicated as intraligand electron transfer (ILET), to emphasize the displacement of charge taking place within the bridging ligand following excitation.

5.2.1.2
$[(bpy)_2Os(tpphz)Os(bpy)_2]^{4+}$

Allowing for the obvious spectral differences, the behavior of $[(bpy)_2Os(tpphz)Os(bpy)_2]^{4+}$ is very similar to that of the ruthenium analogue. In particular, the transient formation of $MLCT_0$ with characteristic absorption at 610 nm is again observed [95]. The behavior is summarized in Fig. 17.

The solvent dependence has the same origin as in the previous case. The shorter time scale for decay to the ground state is as commonly observed for Os(II) relative to Ru(II) complexes, consistent with the energy-gap law.

5.2.1.3
$[(bpy)_2Ru(tpphz)Os(bpy)_2]^{4+}$

In general terms, the behavior of the heteronuclear species $[(bpy)_2Ru(tpphz)Os(bpy)_2]^{4+}$ is similar to that of other Ru(II)–Os(II) polypyridine complexes [22, 83]. Indeed, the conventional photophysics of this complex [91]

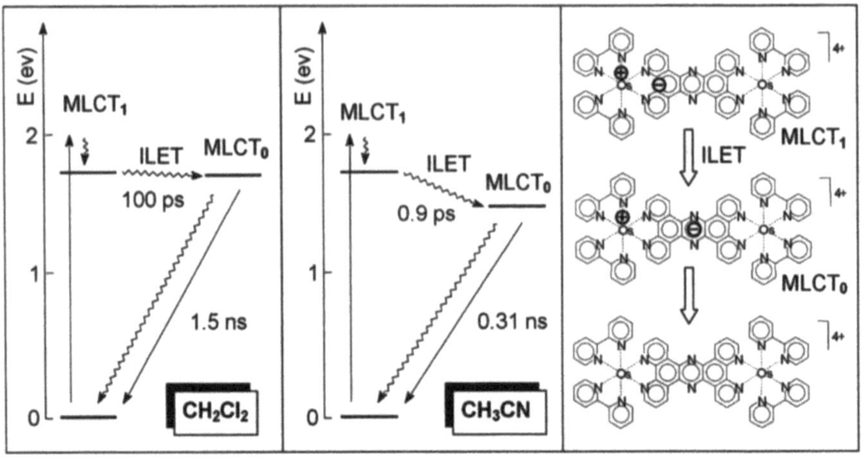

Fig. 17 Photophysical mechanism for $[(bpy)_2Os(tpphz)Os(bpy)_2]^{4+}$ in CH_2Cl_2 (*left*) and CH_3CN (*center*). Schematic picture of the changes in charge distribution along the photophysical pathway (*right*)

reveals the occurrence of efficient (100%) and rapid (< 10 ns) Ru(II) → Os(II) energy transfer. At room temperature, the emission takes place, regardless of excitation wavelength, from the lowest Os-based $MLCT_0$ excited state.

In mechanistic terms, $[(bpy)_2Ru(tpphz)Os(bpy)_2]^{4+}$ is expected to be quite interesting, as the presence of $MLCT_1$ and $MLCT_0$ states on both chromophores provides multiple available pathways for the observed Ru(II) → Os(II) energy transfer process. Two such possible pathways are indicated in Fig. 18: (i) Ru(II) → Os(II) energy transfer at the $MLCT_1$ level (EnT), followed by $MLCT_1$ → $MLCT_0$ relaxation within the Os(II) chromophore (ILET Os); (ii) $MLCT_1$ → $MLCT_0$ relaxation within the Ru(II) chromophore (ILET Ru), followed by Ru(II) → Os(II) energy transfer at the $MLCT_0$ (that, in this schematic representation of the electron distribution, amounts to excited-state metal-to-metal *electron* transfer, MMET).[5] Ultrafast spectroscopy has been used to discriminate between such pathways [95]. The behavior turns out to be strongly dependent on the solvent, not only in terms of kinetics (as was the case of the homonuclear species) but also in terms of mechanism. In fact, the time-resolved spectral changes clearly indicate that pathway (i) is followed in CH_2Cl_2, while pathway (ii) prevails in CH_3CN [95]. The measured rates illustrate the role of the solvent in mechanism selection: in CH_2Cl_2 rapid energy transfer at the $MLCT_1$ level is followed by a slower $MLCT_1$ → $MLCT_0$ relaxation (ILET) in the Os-centered chromophore, whereas in the more polar

[5] A third pathway, involving direct conversion from Ru-based $MLCT_1$ to Os-based $MLCT_0$ is considered less likely.

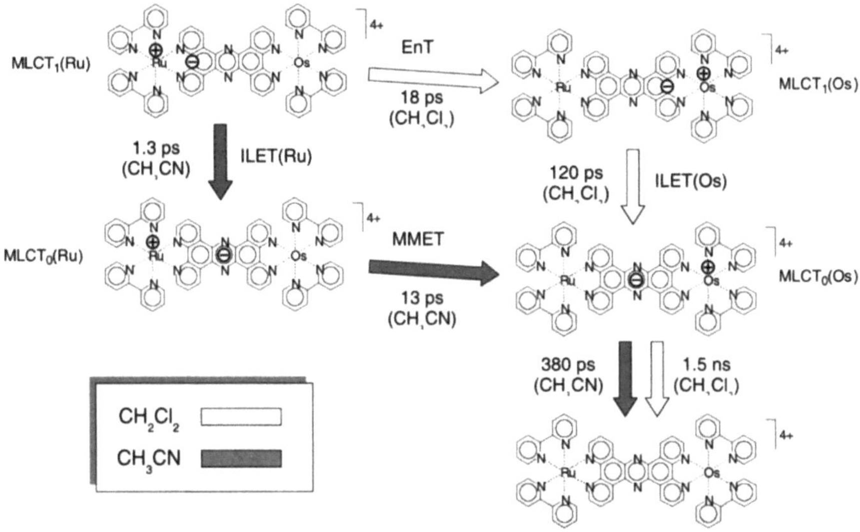

Fig. 18 Mechanistic paths for intercomponent energy transfer in $[(bpy)_2Ru(tpphz)Os(bpy)_2]^{4+}$

CH_3CN ILET in the Ru-centered chromophore is sufficiently fast to precede energy transfer.

The results of this study clearly emphasize how far this bridging ligand is from the standard picture of a simple spacer. In fact, tpphz plays here an active role in mediating the intercomponent energy transfer via its own energy levels. Some of the problems related to the description of the photophysical behavior of bimetallic dyads in conventional terms were mentioned in Sect. 4. The $[(bpy)_2Ru(tpphz)Os(bpy)_2]^{4+}$ complex provides an example of the further intricacies added when the bridging ligand itself has a complex electronic structure. In fact, although the overall process taking place is energy transfer, energy and electron transfer processes are deeply intertwined at the mechanistic level.

5.2.1.4 $[(bpy)_2Ru(tpphz)Os(bpy)_2]^{5+}$

The Ru(II)–Os(III) species $[(bpy)_2Ru(tpphz)Os(bpy)_2]^{5+}$ can be obtained by chemical oxidation of the previously described Ru(II)–Os(II) complex [91]. Upon oxidation the osmium component loses its chromophoric properties and becomes a good electron acceptor. Therefore, as other Ru(II)–Os(III) polypyridine complexes [21] the $[(bpy)_2Ru(tpphz)Os(bpy)_2]^{5+}$ dyad is suitable for the study of photoinduced electron transfer.

The dyad is non-emissive and gives no transient in ns laser photolysis following excitation of the Ru(II) chromophore [91]. The efficient quenching is attributed to electron transfer from the excited Ru(II) chromophore to the Os(III) metal center. It should be noticed that in this system, owing to the presence of $MLCT_1$ and $MLCT_0$ states on the Ru(II) chromophore, the electron transfer quenching could takes place either directly from the $MLCT_1$ or in a stepwise manner, via ILET. Detailed insight into the mechanism can be obtained using ultrafast spectroscopy studies [95]. These studies demonstrate that, in CH_3CN, the quenching of the Ru(II) chromophore proceeds by a sequence of two electron transfer steps (ILET followed by ligand-to-metal electron transfer), as schematically indicated in Fig. 19.

As in the previously discussed Os(II)–Os(II) case, the results again emphasize the key role of the tpphz ligand in the mediating intercomponent transfer processes. The stepwise electron transfer observed in the Ru(II)–Os(III) system clearly does not proceed by superexchange. Rather, it can be considered more close to the hopping mechanism discussed in Sect. 2.2 (although here the bridging ligand is not made of weakly interacting modular subunits). Bridges of this type should be able to sustain charge transfer over long distances. As shown in the next sections, however, the extension towards longer poly-quinoxaline bridges causes further mechanistic complications.

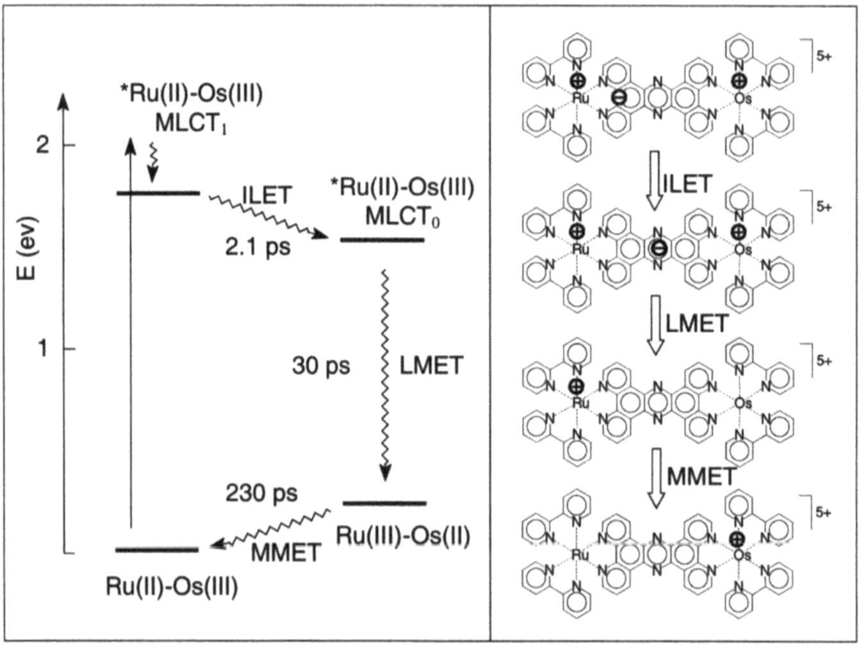

Fig. 19 Photophysical mechanism for $[(bpy)_2Ru(tpphz)Os(bpy)_2]^{5+}$ in CH_3CN (*left*). Schematic picture of the changes in charge distribution along the photophysical pathway (*right*)

5.2.2
tatpp Complexes

Homo-binuclear complexes with the tatpp ligand (**15**) have attracted considerable attention as potential multi-electron catalysts. In particular, $[(phen)_2$-$Ru(tatpp)Ru(phen)_2]^{4+}$ is known to undergo two reversible electrochemical reductions at moderately negative potentials (– 0.18 and – 0.56 V vs. SCE in acetonitrile) [96]. Moreover, it is capable of reversibly storing two electrons (and protons) per molecule upon visible light irradiation in deoxygenated acetonitrile and in the presence of sacrificial reductant agents [97]. In both cases, the doubly reduced form can be considered as a binuclear Ru(II) complex with the tatpp dianion. These interesting features have prompted some detailed photophysical studies on binuclear complexes with the tatpp bridging ligand.

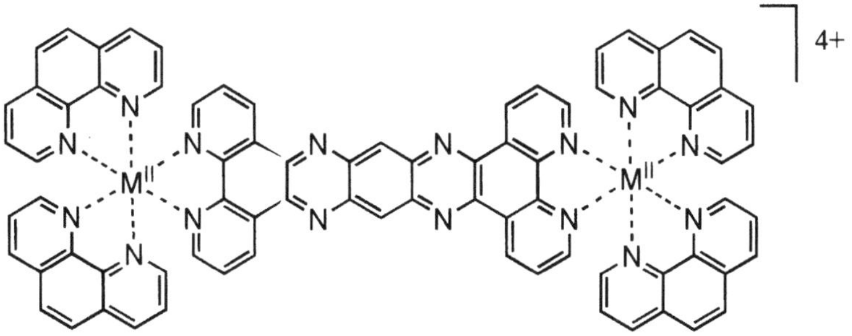

M = Ru, Os

(15)

5.2.2.1
$[(phen)_2Ru(tatpp)Ru(phen)_2]^{4+}$

The absorption spectrum of $[(phen)_2Ru(tatpp)Ru(phen)_2]^{4+}$ exhibits MLCT bands quite similar to those of a $[Ru(phen)_3]^{2+}$ model complex indicating that, again, the spectroscopic transitions lead to higher unoccupied phen-like orbitals of the ligand (Fig. 14) [96]. The complex is appreciably non-emitting. The time-resolved spectroscopic behavior is characterized by the ultrafast appearance (36 ps in CH_3CN, 1 ps in CH_2Cl_2) and slow disappearance (10 ns in CH_3CN, 1 μs in CH_2Cl_2) of a transient absorption with λ_{max} 580 nm. By analogy with the tpphz-based systems, it is tempting to associate this behavior to the population and decay of a lowest MLCT state with the electron localized on the central phenazine-like part of the bridging (see the LUMO in Fig. 14). Such a mechanism was indeed proposed in a preliminary report

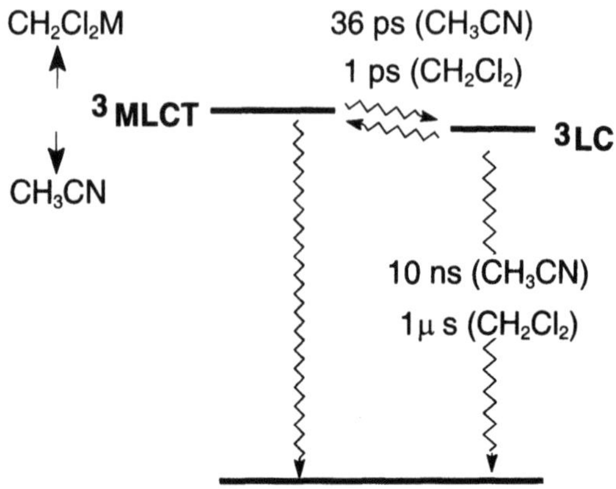

Fig. 20 Photophysical mechanism for $[(phen)_2Ru(tatpp)Ru(phen)_2]^{4+}$

on the time-resolved spectroscopy of this complex [98]. On closer examination, however, the behavior of $[(phen)_2Ru(tatpp)Ru(phen)_2]^{4+}$ proves to be quite different from that of the tpphz analogue. In fact, because of the high degree of conjugation of the tatpp ligand, the ligand-centered (LC) π–π^* triplet is lowered considerably, to an energy close or even lower than the MLCT state. Definite proof that the 580-nm transient is indeed ligand-centered (LC) π–π^* triplet is obtained from the transient spectroscopy of the tatpp ligand (as a Zn complex) [99]. Thus, a plausible mechanism for the photophysical behavior of $[(phen)_2Ru(tatpp)Ru(phen)_2]^{4+}$ is sketched in Fig. 20.

The strong solvent dependence of formation and decay of the ^{3}LC state can be easily explained in terms of a solvent-sensitive equilibrium between the intrinsically long-lived, non-polar LC state and an intrinsically short-lived, polar MLCT state (similar situations can be found in bichromophoric systems such as, e.g., Ru(polypyridine)/pyrene dyads) [100].

5.2.2.2 $[(phen)_2Os(tatpp)Os(phen)_2]^{4+}$

In the case of the tpphz bridging ligand, ruthenium and osmium homonuclear complexes exhibited qualitatively the same behavior (Sects. 5.2.1.1 and 5.2.1.2). By contrast, the photophysics of $[(phen)_2Os(tatpp)Os(phen)_2]^{4+}$ is completely different from that of the ruthenium analogue. In fact, the transient behavior of $[(phen)_2Os(tatpp)Os(phen)_2]^{4+}$ is characterized by an ultrafast *depletion* of absorption at 580 nm (0.4 ps in CH_3CN, 3.8 ps in CH_2Cl_2), followed by a slower decay of the resulting spectrum to the original baseline (3.9 ps in CH_3CN, 50 ps in CH_2Cl_2) [99]. The difference in behavior is associated to a difference in the energy levels of the system. In the osmium complex, because of the easier oxidation of the metal, the whole MLCT excited-state

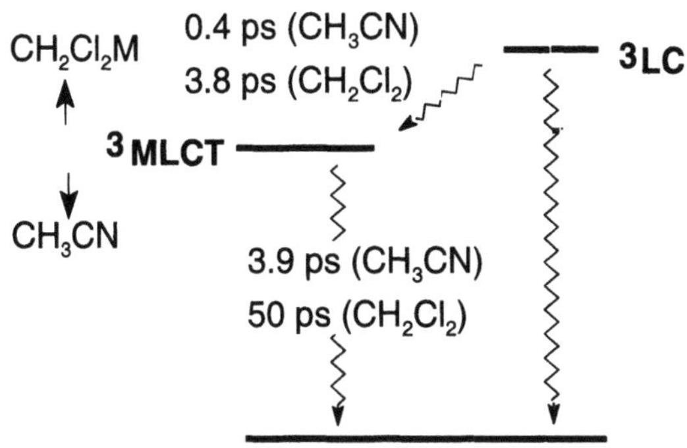

Fig. 21 Photophysical mechanism for $[(phen)_2Os(tatpp)Os(phen)_2]^{4+}$

manifold is expected to be lower by ca. 0.5 eV with respect to the analogous ruthenium system. As a consequence, the relative energies of the lowest excited states of the system are likely to be reversed, with the MLCT triplet being lower than the LC one. The situation is sketched in Fig. 21. The ultrafast decay of the MLCT state is in the ps time scale, as expected for a low-energy MLCT state, and its solvent dependence can be related to energy-gap-law effects [99].

5.2.3 bqpy Complexes

The bqpy ligand can be viewed as an analogue of tatpp with a pyrene unit substituting the central benzene ring. With a more extended π system, bqpy complexes could be expected as a further step towards delocalized behavior, characterized by facile reduction and low-lying MLCT and LC states. In fact, the behavior of $[(bpy)_2Ru(bqpy)Ru(bpy)_2]^{4+}$ **(16)** deviates sharply from these expectations [101]. For instance, one-electron reduction (– 0.90 V vs. SCE) is much more difficult than for the analogous tatpp (– 0.18 V vs. SCE) and even than tpphz (– 0.71 V vs. SCE) complexes. Also, the photophysical properties (intense, long-lived MLCT emission at 628 nm in CH_3CN) are quite different

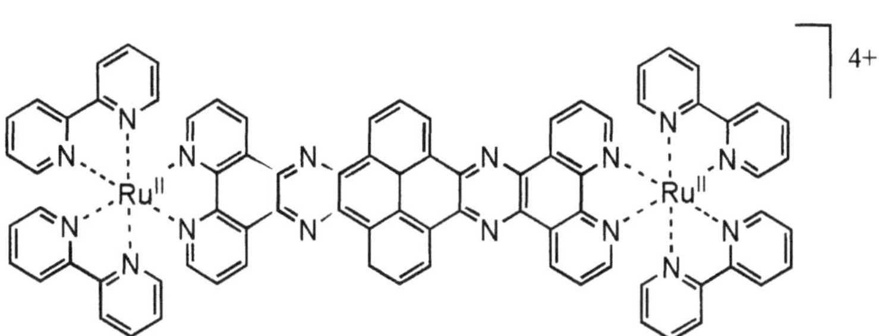

(16)

from those of tatpp (no emission, lowest LC excited state) and tpphz (weak, short-lived MLCT emission at 740 nm in CH_3CN). The anomalous behavior of bqpy complexes with respect to the other ladder-type systems can be related to the presence of additional angularly fused rings. This causes the LUMO of the ligand to have a peculiar nodal structure (nodes both at the pyrazine nitrogens and at central carbons) and a very high energy (Fig. 14). Although at the moment no bqpy heteronuclear complexes are available, it would be interesting to investigate how these peculiar properties may affect intercomponent electron/energy processes.

6 Conclusions

Photoinduced electron and energy transfer processes in binuclear metal complexes have been discussed, with particular emphasis on the role of the molecular bridges connecting the donor and acceptor components. The binuclear complexes considered have metal polypyridine components, with Ru(II) units acting as photoexcited electron/energy donors, Os(II) units as energy acceptors, and Rh(III) and Os(III) units as electron acceptors. Two classes of rigid bridges have been considered, based on polyphenylene and polyquinoxaline structures. The two types of bridge behave rather differently with regard to their electron/energy transfer mediating ability.

The polyphenylene bridges perform, both in energy and electron transfer, as well-behaved superexchange mediators, with exponential dependence of transfer rate on bridge length. The experimental attenuation factors β in the range 0.34–0.65 Å^{-1}, permit efficient energy and electron transfer over long metal-metal distances, e.g., 42 Å for energy transfer in Ru(II)–Os(II), 24 Å for electron transfer in Ru(II)–Rh(III). The results demonstrate that with this type of bridges, the twist angle between adjacent phenylene units is an extremely important parameter. In fact, alkyl substitution in a single unit of the polyphenylene chain causes a strong decrease of energy and electron transfer rate constants. This effect can be easily rationalized considering the increase in the twist angle accompanying substitution and the concomitant decrease in the electronic coupling between adjacent modules of the bridge. The sensitivity of transfer rates to twist angles could, in principle, be exploited to achieve tuning or switching of signals along polyphenylene molecular wires.

The polyquinoxaline bridges, because of their fully aromatic structure, contribute low-lying electronic levels to the dyads. As a consequence, they are likely to participate in the energy/electron transfer processes in a more direct way. Indeed, the shortest member of the series, tpphz, mediates energy transfer in the Ru(II)–Os(II) system by a solvent dependent sequence of steps involving the unoccupied orbitals of the bridge. Also, photoinduced electron transfer in the Ru(II)–Os(III) system proceeds stepwise through the bridging ligand, with some similarities to a charge injection/hopping mechanism. The delocalized nature of these bridges causes further problems when the length of the bridges is increased. In fact, complexes of the next member in the series, tatpp, have not only (as expected) MLCT states of lower energy, but also ligand-centered (LC) triplet states at very low energy. Thus, e.g., in Os(II) tatpp complexes the situation is similar to that of analogous tpphz systems (lowest MLCT state), but for Ru(II) complexes the lowest excited state is LC. For long bridges, such low-energy LC states are likely to act as an energy sink, preventing the study of intercomponent transfer processes. It should be pointed out, however, that these systems may also hide some surprise. In fact, the longest member of the series, bqpy, behaves in many respects as the shortest tpphz, with high-energy redox levels and MLCT states. This is apparently related to the presence of angularly fused rings, which strongly affect the nodal properties of the molecular orbitals. These effects could probably be used in the design of longer ladder-type bridging ligands without the handicap of low-lying LC states.

Acknowledgements We thank Luisa De Cola for providing relevant results prior to publication. Financial support from the EC (G5RD-CT-2002-00776, MWFM) and MIUR (FIRB-RBNE019H9K) is gratefully acknowledged.

References

1. Jortner J, Ratner M (eds) (1997) Molecular Electronics. Blackwell, London
2. Aviram A, Ratner MA (eds) (1998) Special volume on molecular electronics. Ann N Y Acad Sci 852
3. Joachim C, Gimzewski JK, Aviram A (2000) Nature 408:541
4. Tour JM (2000) Acc Chem Res 33:791
5. Pease AR, Jeppesen JO, Stoddart JF, Luo Y, Collier CP, Heath JR (2001) Acc Chem Res 34:433
6. Weber K, Creager SE (1994) Anal Chem 66:3164
7. Creager S, Yu CJ, Bamad C, O'Connor S, MacLean T, Lam E, Chong Y, Olsen GT, Luo J, Gozin M, Kayyem JF (1999) J Am Chem Soc 121:1059
8. Sachs SB, Dudek SP, Hsung RP, Sita LR, Smalley JF, Newton MD, Feldberg SW, Chidsey CED (1997) J Am Chem Soc 119:10 563
9. Sykes HD, Smalley JF, Dudek SP, Cook AR, Newton MD, Chidsey CED, Felberg SW (2001) Science 291:1519
10. Chidsey CED (1991) Science 251:919
11. Donhauser ZJ, Mantooth BA, Kelly KF, Bumm LA, Monnell JD, Stapleton JJ, Price Jr DW, Rawlett AM, Allara DL, Tour JM, Weiss PS (2001) Science 292:2303
12. Reed MA, Zhou C, Muller CJ, Burgin TP, Tour JM (1997) Science 278:252
13. Chen J, Reed MA, Rawlett AM, Tour JM (1999) Science 286:1550
14. Tour JM, Rawlett AM, Kozaki M, Yao Y, Jagessar RC, Dirk SM, Price DW, Reed MA, Zhou C-W, Chen J, Wang W, Campbell I (2001) Chem Eur J 7:5118

15. Leathermann G, Durantini EN, Gust D, Moore TA, Moore AL, Stone S, Zhou Z, Rez P, Liu YZ, Lindsay SM (1999) J Phys Chem B 103:4006
16. Haag R, Rampi MA, Holmlin RE, Whitesides GM (1999) J Am Chem Soc 121:7895
17. Holmlin EH, Ismagilov RF, Haag R, Mujica V, Ratner MA, Rampi MA, Whitesides GM (2001) Angew Chem Int Ed 40:2316
18. Andreas RP, Bein T, Dorogi M, Feng S, Henderson JI, Kubiak CP, Mahoney W, Osif RG, Reifenberger R (1996) Science 272:1323
19. Cui XD, Primak A, Zarate X, Tomfohr J, Sankey OF, Moore AL, Moore TA, Gust D, Harris G, Lindsay SM (2001) Science 294:571
20. Paddon-Row MN (2001) Covalently linked systems based on organic components. In: Balzani V (ed) Electron transfer in chemistry, vol III. Wiley, Weinheim, Chap 2.3, p 179
21. Scandola F, Chiorboli C, Indelli MT, Rampi MA (2001) Covalently linked systems containing metal complexes. In: Balzani V (ed) Electron transfer in chemistry, vol III. Wiley, Weinheim, Chap 2.1, p 337
22. De Cola L, Belser P (2001) Photonic wires containing metal complexes. In: Balzani V (ed) Electron transfer in chemistry, vol V. Wiley, Weinheim, Chap 3, p 97
23. Davis WB, Svec WA, Ratner MA, Wasielewski MR (1998) Nature 396:60
24. Marcus RA (1964) Annu Rev Phys Chem 15:155
25. Sutin N (1983) Prog Inorg Chem 30:441
26. Marcus RA, Sutin N (1985) Biochim Biophys Acta 811:265
27. Jortner J (1976) J Chem Phys 64:4860
28. Ulstrup J (1979) Charge transfer processes in condensed media. Springer, Berlin Heidelberg New York
29. Miller JR, Beitz JV, Huddleston RK (1984) J Am Chem Soc 106:5057
30. Meyer TJ, Taube H (1987) In: Wilkinson SJ, Gillard RD, McCleverty JA (eds) Comprehensive coordination chemistry, vol 1. Pergamon Press, Oxford, p 331
31. Newton MD (1991) Chem Rev 91:767
32. Jortner J, Bixon M, Langenbacher T, Michel-Beyerle ME (1998) Proc Natl Acad Sci USA 95:12 759
33. Oevering, H, Verhoeven JW, Paddon-Row MN, Warman JM (1989) Tetrahedron 45:4751
34. Halpern J, Orgel LE (1960) Disc Faraday Soc 29:32
35. McConnell HM (1961) J Chem Phys 35:508
36. Mayoh B, Day P (1974) J Chem Soc Dalton:846
37. Miller JR, Beitz JV (1981) J Chem Phys 74:6746
38. Richardson DE, Taube H (1983) J Am Chem Soc 10:540
39. Scandola F, Argazzi R, Bignozzi CA, Chiorboli C, Indelli MT, Rampi MA (1993) Coord Chem Rev 125:283
40. Meggers E, Michel-Beyerle ME, Giese B (1998) J Am Chem Soc 120:12 950
41. Davis WB, Svec WA, Ratner MA, Wasielewski MR (1998) Nature 396:60
42. Lewis FD (2001) Electron transfer and charge transport processes. In: Balzani V (ed) Electron transfer in chemistry, vol III. Wiley, Weinheim, Chap 1.5, p 105
43. Weiss EA, Ahrens MJ, Sinks LE, Gusev AV, Ratner MA, Wasielewsi MR (2004) J Am Chem Soc 126:5577
44. Orlandi G, Monti S, Barigelletti F, Balzani V (1980) Chem Phys 52:313
45. Murtaza Z, Zipp AP, World LA, Graff D, Jones WE Jr, Bates WD, Meyer TJ (1991) J Am Chem Soc 113:5113
46. Naqvi KR, Steel C (1993) Spectrosc Lett 26:1761
47. Balzani V, Bolletta F, Scandola F (1980) J Am Chem Soc 102:2152

48. Lamola AA (1969) In: Lamola AA, Turro NJ (eds) Energy transfer and organic photochemistry. Wiley, New York
49. Turro NJ (1978) Modern molecular photochemistry. Benjamin Cummings, Menlo Park
50. Crosby GA (1983) J Chem Educ 60:791
51. Scandola F, Balzani V (1983) J Chem Educ 60:814
52. Scholes GD, Ghiggino KP, Oliver AM, Paddon-Row MN (1993) J Phys Chem 97:11 871
53. Oevering H, Verhoeven JW, Paddon-Row MN, Cotsaris E, Hush NS (1988) Chem Phys Lett 143:488
54. Closs GL, Piotrowiak P, McInnis JM, Fleming GR (1988) J Am Chem Soc 110:2652
55. Closs GL, Johnson MD, Miller JR, Piotrowiak P (1989) J Am Chem Soc 111:3751
56. Lehn J-M (1988) Angew Chem Int Ed Engl 27:89
57. Cram DJ (1988) Angew Chem Int Ed Engl 27:1009
58. Pedersen CJ (1988) Angew Chem Int Ed Engl 27:1021
59. Balzani V, Scandola F (1991) Supramolecular photochemistry. Horwood, Chichester, UK
60. Lehn J-M (1995) Supramolecular chemistry: concepts and perspectives. Wiley VCH, Weinheim
61. Juris A, Balzani V, Barigelletti F, Campagna S, Belser P, Von Zelewsky A (1988) Coord Chem Rev 84:85
62. Kalyanasundaram K (1992) Photochemistry of polypyridine and porphyrin complexes. Academic, New York
63. Heath GA, Yellowlees LJ, Braterman PS (1982) Chem Phys Lett 92:646
64. Chang YJ, Xu X, Yabe T, Yu S-C, Anderson, DR, Orman LK, Hopkins JB (1990) J Phys Chem 94:729
65. Barigelletti F, Flamigni L (2000) Chem Soc Rev 29:1
66. Sauvage J-P, Collin J-P, Chambron J-C, Guillerez S, Coudret C, Balzani V, Barigelletti F, De Cola L, Flamigni L (1994) Chem Rev 94:993
67. Liang YY, Baba AI, Kim WY, Atherton SJ, Schmehl RH (1996) J Phys Chem 100:18 408
68. Indelli MT, Scandola F, Collin J-P, Sauvage J-P, Sour A (1996) Inorg Chem 35:303
69. Indelli MT, Scandola F, Flamigni L, Collin J-P, Sauvage J-P, Sour A (1997) Inorg Chem 36:4247
70. Collin J-P, Gavina P, Hietz V, Sauvage J-P (1998) Eur J Inorg Chem :1
71. Barigelletti F, Flamigni L, Collin J-P, Sauvage J-P (1997) Chem Commun:333
72. Barigelletti F, Flamigni L, Guardigli M, Juris A, Beley M, Chodorowski-Kimmes S, Collin J-P, Sauvage J-P (1996) Inorg Chem 35:136
73. Constable EC, Cargill Thompson AM W (1996) New J Chem 20:65
74. Schlicke B, Belser P, DeCola L, Sabbioni E, Balzani V (1999) J Am Chem Soc 121:4207
75. Welter S, Salluce N, Belser P, Groeneveld M, De Cola L Coord Chem Rev (in press)
76. Indelli MT, Bignozzi CA, Harriman A, Schoonover JR, Scandola F (1994) J Am Chem Soc 116:3768
77. Helms A, Heiler D, McLendon G (1992) J Am Chem Soc 114:6227
78. Helms A, Heiler D, McLendon G (1991) J Am Chem Soc 113:4325
79. Tour JM, Lamba JJS (1993) J Am Chem Soc 115:4935
80. Tsuzuki K, Tanabe J (1991) J Phys Chem 95:139
81. Verhoeven JW, Paddon-Row MN, Warman JM (1992) In: Kochanski E (ed) Photoprocesses in transition metal complexes: biosystems and other molecules, experiment and theory. Kluwer, Dordrecht, p 271
82. Johnson MD, Miller JR, Green NS, Closs GL (1989) J Phys Chem 93:1173

83. Joran AD, Leland BA, Felker PM, Zewail AH, Hopfield JJ, Dervan PD (1987) Nature 327:508
84. Clayton AH, Ghiggino KP, Wilson GJ, Keyte PJ, Paddon-Row MN (1992) Chem Phys Lett 195:249
85. Helms A, Heiler D, McLendon G (1992) J Am Chem Soc 114:6227
86. De Cola L, Belser P (1998) Coord Chem Rev 177:301
87. Harriman A, Khatyr A, Ziessel R, Benniston A (2000) Angew Chem Int Ed 39:4287
88. El-ghayyoury A, Harriman A, Khatyr A, Ziessel R (2000) Angew Chem Int Ed 39:185
89. Ziessel R, Hissler M, El-ghayoury A, Harriman A (1998) Coord Chem Rev 177:1251
90. Bolger J, Gourdon A, Ishow E, Launay JP (1996) Inorg Chem 35:2937
91. Chiorboli C, Bignozzi CA, Scandola F, Ishow E, Gourdon A, Launay JP (1999) Inorg Chem 38:2402
92. Campagna S, Serroni S, Bodige S, MacDonnell FM (1999) Inorg Chem 38:692
93. Olson EJC, Hu D, Hörmann A, Jonkmann AM, Arkin MR, Stemp EDA, Barton JK, Barbara PF (1997) J Am Chem Soc 119:11 458
94. Flamigni L, Encinas S, Barigelletti F, MacDonnell FM, Kim M-J, Puntoriero F, Campagna S (2000) Chem Commun:1185
95. Chiorboli C, Rodgers MA J, Scandola F (2003) J Am Chem Soc, 125:483
96. Kim M-J, Konduri R, Ye H, MacDonnell FM, Puntoriero F, Serroni S, Campagna S, Holder T, Kinsel G, Rajeshwar K (2002) Inorg Chem 41:2471
97. Konduri R, Ye H, MacDonnell FM, Serroni S, Campagna S, Rajeshwar K (2002) Angew Chem Int Ed 41:3185
98. Chiorboli C, Fracasso S, Scandola F, Campagna S, Serroni S, Konduri R, MacDonnell FM (2003) Chem Commun:1658
99. Chiorboli C, Fracasso S, Ravaglia M, Standola F, Campagna S, Wouters KL, Konduri R, MacDonnell FM, Inorg Chem (in press)
100. Indelli MT, Ghirotti M, Prodi A, Chiorboli C, Scandola F, McClenaghan ND, Puntoriero F, Campagna S (2003) Inorg Chem 42:5489
101. Ishow E, Gourdon A, Launay J-P, Chiorboli C, Scandola F (1999) Inorg Chem 38:1504
102. Dresselhaus MS, Dresselhaus G, Avouris P (2000) Carbon Nanotubes. Springer-Verlag
103. Bachtold A, Hadley P, Nakanishi T, Dekker C (2001) Science 294:1317
104. Huang Y, Duan X, Cui Y, Lauhon LJ, Kim K, Lieber CM (2001) Science 294:1313

Top Curr Chem (2005) 257: 103–133
DOI 10.1007/b136068

Published online: 6 July 2005

Molecules as Wires: Molecule-Assisted Movement of Charge and Energy

Emily A. Weiss · Michael R. Wasielewski (✉) · Mark A. Ratner

Department of Chemistry and Center for Nanofabrication and Molecular Self-Assembly, Northwestern University, 2145, N. Sheridan Rd., Evanston, IL 60208-3113, USA
eweiss@chem.northwestern.edu, wasielew@chem.northwestern.edu, ratner@chem.northwestern.edu

1	**Introduction**	104
2	**Mechanisms of Charge Transfer**	106
2.1	Superexchange	107
2.2	Sequential Charge Transfer	107
3	**Factors that Determine the Charge Transfer Mechanism**	109
3.1	Electronic Coupling	109
3.1.1	Calculation of Electronic Coupling	110
3.1.2	Estimation of Donor–Acceptor Coupling via Magnetic Exchange Measurements	111
3.2	Energy Matching	115
4	**Methods of Calculating Conductance**	117
5	**Examples of Molecular Wire Systems**	119
5.1	Oligo(phenylenevinylene)	119
5.2	Oligophenylene	121
5.3	DNA	123
5.4	Oligothiophene	124
5.5	Photonic Wires	126
6	**Remarks**	126
	References	128

Abstract In this chapter, we explore experimental and theoretical aspects of molecular wire-like charge transport from the mechanistic point of view. We discuss competition between coherent superexchange and sequential mechanisms of transport through donor-bridge-acceptor systems, where the donor and acceptor are either molecules or metal/semiconductor contacts. The focus is on the two major determinants of mechanism: electronic coupling and energy level matching. Some methods of calculating conductance are outlined, and the relationship between conductance in a metal/molecule/metal junction and electron transfer, where donor and acceptor have relatively discrete electronic energy levels is explored. Finally, we give several examples of chemical systems that have displayed wire-like behavior and discuss their characterization.

Keywords Molecular wire · Superexchange · Electron transfer · Energy Transfer · Electronic coupling

1
Introduction

The term "molecular wire" has been used liberally in chemistry, physics, and materials science literature. In some cases, it describes a system with a very specific behavior; in others, it refers simply to the shape of the molecule under consideration. A molecular wire has been defined as "a molecule connected between two reservoirs of electrons" [1], more stringently as "a molecule that conduct(s) electrical current between two electrodes" [2], and even more narrowly as a device that conducts in a regime "wherein the distance dependence [of electron transfer (ET)] may be very weak" [3].

More important than the nomenclature is the desired function of a molecular wire system, namely, robust and directional transport of charge or of energy. Are molecules the ultimate interchangeable electronic components? Because they can be tuned by an infinite series of synthetic adjustments, they offer a potential for controllability and versatility superior to that of metals and semimetals. Defined orbitals of predisposed shape dictate where within the molecule the charge carrier can move, and with what energy. An even more subtle property of molecules is that their electronic distribution may decide whether the spin properties of the traveling electron will be conserved during the transfer or redistributed in the form of spin polarizing to other electrons in the system.

Although energy transport along a molecular wire might be thought of as motion of an electron/hole pair or Frenkel exciton (a boson) and charge transfer (CT) might be seen as motion of one hole or one electron (a fermion), they are closely related phenomena. From a transport viewpoint, the important insight is that molecular wires should behave as Hückel-like or tight-binding systems, dominated by nearest-neighbor contacts. Then we should be able to write, quite generally,

$$
\begin{aligned}
H &= H_{\mathrm{P}} + H_{\mathrm{E}} + V_{\mathrm{PE}} \qquad (1)\\
H_{\mathrm{P}} &= \sum_i c_i^+ c_i \varepsilon_i + \sum_{ij} V_{ij}(c_i^+ c_j + c_j^+ c_i)\\
H_{\mathrm{E}} &= \sum_\lambda \left[\left(b_\lambda^+ b_\lambda + \frac{1}{2}\right)\hbar\omega_\lambda + anh\right]\\
V_{\mathrm{PE}} &= \sum_{ij}\sum_\lambda M_\lambda^{ij} c_i^+ c_j (b_\lambda^+ + b_\lambda)
\end{aligned}
$$

Here, H_{P}, H_{E}, and V_{PE} are the Hamiltonians for the moving particle, the environment, and the particle/environment interaction respectively. The operators b_λ^+ and c_i^+ create, respectively, environment boson excitations (phonons, vibrations, photons, solvent modes) and carriers (electrons or excitons). The environment bosons have index λ and frequency ω_λ, the particles have site energy ε_i and tunneling matrix elements V_{ij}; the sum in the second term of H_P is usually restricted to nearest neighbors or next-nearest neighbors. This restriction is physical and fundamental, arising from the molecular organization of space. The term denoted (anh) corresponds to the anharmonic behavior of the environmental bosons, and is almost always ignored.

While many significant terms such as the electronic repulsion and exchange terms and all spin-dependent structure are missing from Eq. 1, much of what is needed to understand molecular wire behavior is present; indeed Eq. 1 contains the Marcus–Jortner–Dogonadze Hamiltonians for intramolecular electron transfer and the Fröhlich—Toyozawa–Holstein polaron treatment as special cases.

The operator c_i^+ can be of boson type (for exciton-mediated energy transfer) or of Fermion type (for hole- or electron-mediated CT). The energy quantity M_λ^{ij} characterizes the coupling involved when changing either population $(i = j)$ or motion $(i \neq j)$ of the carrier.

Since excitons are electron/hole pairs, it is reasonable to expect that the tunneling matrix element for excitons might be proportional to a product of matrix elements for hole and electron tunneling. This is discussed in Sect. 3.1.2.

A final note should be stressed concerning the Hamiltonian of Eq. 1. The local sites denoted (i, j) might be atomic orbitals or molecular orbitals of a fragment. For example, in a p-phenylene wire, they might be the $p\pi$ orbitals on the individual carbons or the frontier orbitals on each C_6H_4 fragment. Each of these representations is useful. The latter, however, is most often used, essentially because otherwise the matrix elements V_{ij} are like those in actual Hückel theory; those values are on the order of 2–3 eV, and then the assumption of relatively weak electronic coupling, nearly always assumed in transport, becomes invalid. For quantitative study Eq. 1 is usually replaced either by the full electronic Coulomb Hamiltonian or by semi-empirical atom-based Hamiltonians such as CNDO, INDO, or PPP [4–6].

In this chapter, we will discuss both mechanisms and experimental evidence for wire-like transport (simply meaning transport arising from motion assisted by the molecular bridge) of electronic charge (in donor/bridge/acceptor intramolecular ET rate constants and in molecular electronic transport junction conductance) and of energy. The focus is on behaviors; because this field is so rich and the phenomena are so important, the literature is vast. The chapter is intended as a sketch, not as a review or a portrait.

In keeping with the theme of molecules as individual wires, we will not deal at all with the closely related and even larger field of charge and energy transport in molecular materials. Many of the mechanisms (coherent tunneling, thermal hopping, Förster transfer, Dexter exchange energy flow, defect-dominated transport) occur both in single-wire systems and in molecular materials such as conductive polymers and molecular crystals.

Section 2 outlines the different mechanisms for CT, while Section 3 discusses the matrix elements and energy differences that help determine the mechanism. After a very brief discussion of conductance calculations in Section 4, Section 5 is devoted to examples of molecular wire systems.

2 Mechanisms of Charge Transfer

Long-distance CT is intrinsically a nonadiabatic process [7, 8] in which the rate of CT is dictated by some combination of a strongly distance-dependent tunneling mechanism and weakly distance-dependent incoherent transport. The tunneling mechanism is often considered to be that of superexchange, the transfer of electrons or holes from donor to acceptor through an energetically well-isolated bridge, during which bridge orbitals are utilized solely as a coupling medium with no nuclear motion along the bridge [9–11]. Incoherent or sequential CT involves real intermediate states that couple to internal nuclear motions of the bridge and the surrounding medium, and are therefore energetically accessible and change their geometry [12]. This is called a (thermally activated) hopping mechanism [3, 13].

The β parameter, the characteristic parameter for the decay of CT rate constant k as a function of distance, r_{DA},

$$k = k_0 \mathrm{e}^{-\beta r_{\mathrm{DA}}} \tag{2}$$

has been used as a benchmark for the quality of a molecular wire and, in fact, to determine whether a donor–bridge–acceptor system should be considered to display wire-like behavior at all. The smaller the β value, the longer the distance over which charge can be transferred without penalty. The β parameter is often used to characterize all types of transport, although by definition it only applies to exponentially decaying processes. Some may liken the limit of small β to band transport, a so-called π-way through which the electron can travel coherently [14]. This can be confusing because shallow distance dependence can also be achieved via a series of short tunneling events, the incoherent hopping mechanism, which should not lead to exponential decay with length. Additionally, β should be considered intrinsic not to the molecule serving as the bridge, but rather to the wire system as a whole, i.e., donor+bridge+acceptor, whether the donor and acceptor are molecular species or metallic contacts.

The range of β values found for identical bridge units measured in solution and through self-assembled monolayers (SAMs) of organic thiols on the surfaces of metal electrodes (Ag, Au, and Hg) reflects the fact that the most fundamental aspects of transport, including length dependence, are sensitive to environment (e.g., $\beta = 0.6-1.2$ Å^{-1} for alkanes [15–23], $\beta = 0.32-0.66$ Å^{-1} for oligophenylenes [24–31], $\beta = 0.06-0.5$ Å^{-1} for oligo-(phenylene ethynylene)s (OPEs) and oligo(phenylenevinylene)s (OPVs) [32–34], $\beta = 0.04-0.2$ Å^{-1} for oligoenes [35–37], and $\beta = 0.04-0.17$ Å^{-1} for oligoynes [38–40]).

2.1 Superexchange

Superexchange is thought to be an important mechanism for efficient ET within the photosynthetic reaction center [41, 42] and has been studied in various biomimetic systems [43, 44]. The term was first used by Kramers [9] and later by Anderson [10, 11] to describe the indirect exchange coupling of unpaired spins via orbitals having paired spins, which acquire paramagnetic character through mixing with CT excited state configurations [11]. In CT occurring via superexchange, no charge ever actually resides on the bridge, and the states that the molecule occupies between the time when the electron leaves the donor and when it arrives at the acceptor are called virtual excitations. The quantity that dictates the probability of transmission of an electron from donor to acceptor in this way is the electronic superexchange coupling, t_{DA} [45, 46]. McConnell gives a perturbation-theory-based expression for the magnitude of the donor–acceptor superexchange coupling, t_{DA}, in terms of individual resonance integrals between molecular subunits and the energy gap between the degenerate donor and acceptor and the homologous bridge [47]. The McConnell model predicts an approximately exponential dependence of t_{DA} on the donor–acceptor distance, r_{DA}, with decay parameter β^{SE}:

$$\beta^{SE} = -\,2(\ln|t/\Delta|)/r\,, \tag{3}$$

where t is the nearest-neighbor transfer integral, Δ is the energy of the bridge orbitals relative to the donor/acceptor, and r is the width of one subunit. This approximation is verified by many experimental data [48–50]. Significant extensions of this model have been derived to accommodate non-nearest-neighbor interactions [51], multiple geometry-dependent CT pathways [52, 53] and a nonhomologous bridge [54].

2.2 Sequential Charge Transfer

As the length of a bridge increases, the superexchange interaction (decaying exponentially) can become small. In this case, if donor and bridge are within several k_BT of resonance, the rate constant for CT may be dominated by the incoherent term. According to Jortner et al. [55], sequential CT only occurs when there is near-resonant charge injection, vibronic overlap of the ion pair states formed as the charge moves from bridge site to bridge site, and vibronic overlap of the ion pair state in which the charge is located on the terminal bridge site with that in which it is located on the acceptor. The distance dependence of hopping transport is sensitive to the nature of the diffusion of the charge from donor to acceptor. For a reaction with sufficient free energy for the charge to get to the acceptor, the CT rate is given by [14, 55, 56]:

$$\ln k_{\mathrm{CT}} \propto - \eta \ln N \,, \tag{4}$$

where η is between 1 and 2, N is the number of hopping steps, and the proportionality constant is dependent on the individual tunneling probabilities. For long distances, the incoherent, wire-like channel generally becomes more efficient than the coherent one [3, 57].

Both analytical [58] and numerical density matrix [59, 60] approaches have been used to understand the competition between superexchange and sequential mechanisms for the limiting case of three states. In the initial state, donor, bridge, and acceptor are neutral. In the intermediate state, whose energy is systematically varied, the donor is oxidized and the bridge is reduced. In the final state, the donor is oxidized and the acceptor is reduced. These studies predict that, in a regime in which there is a large energy gap between the initial and intermediate states, superexchange is effective because it is an activationless mechanism. However, if this energy gap is comparable to the reorganization energy or electronic coupling, superexchange and sequential mechanisms can compete. Tang [61] has derived a general expression for CT rate that includes activationless and activated terms and reflects this competition. The result has also been generalized to disordered extended systems [62].

For understanding the mechanistic transfer from coherent tunneling [found for big injection gaps (alkanes) and low temperatures] to incoherent hopping [found for small gaps (π-systems) and higher temperatures], it is useful to think in terms both of energies and of times. Figure 1 shows the situation schematically: the bridge sites (here assumed degenerate) couple according to the H in Eq. 1. Figure 1a and b are drawn for the local bridge frontier orbitals and for the bridge eigenstates, respectively: the bandwidth W varies from $2V_{\mathrm{B1B2}}$ for a two-site bridge to $4V_{\mathrm{B1B2}}$ for an infinite length bridge. The value E_{G} is the injection gap energy (here shown for electron injection, but the hole injection is very similar). The levels labeled **D**, **A** can be the donor and acceptor frontier orbital levels (for ET kinetics) or the Fermi levels (for conductance junctions).

If $k_{\mathrm{B}}T$ is of order E_{G} or $E_{\mathrm{G}} - W/2$, then thermal injection is relatively facile, and we expect to observe either resonant tunneling (for weak vibronic coupling—small M_{λ}^{ij} of Eq. 1—as in carbon nanotube wires) or localized hopping, if the vibronic coupling is stronger.

The second consideration involves a generalization of the Landauer/Buttiker contact time for bridge tunneling [63]. This time, τ_{LB}, is the time that the electron is actually on the bridge. In the limit of large E_{G}, it is of the simple

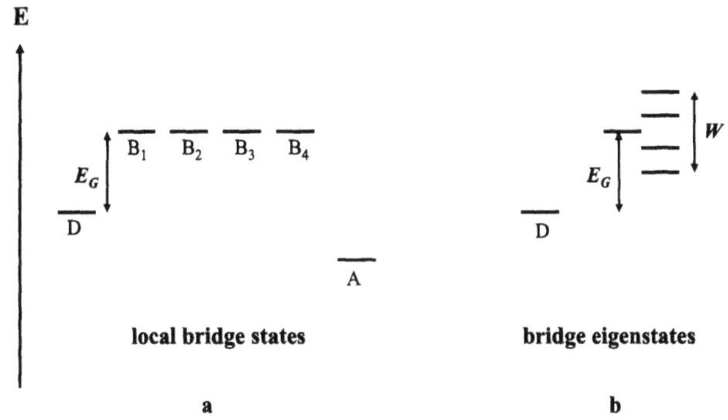

Fig. 1 Degenerate bridge sites in two representations: Panel **a** is drawn for the local bridge frontier orbitals and **b** is drawn for the bridge eigenstates. Bandwidth W varies from $2V_{\mathrm{B1B2}}$ for a two-site bridge to $4V_{\mathrm{B1B2}}$ for an infinite length bridge. The value E_G is the injection gap energy (here shown for electron injection, but the hole injection is very similar). The levels labeled D and A can be the donor and acceptor frontier levels (for electron transfer kinetics) or the Fermi levels (for conductance junctions)

uncertainty principle form

$$\tau_{\mathrm{LB}} \approx \frac{N\hbar}{E_\mathrm{G}} \tag{5}$$

with N the number of bridge sites. If E_G is large and N is small (alkane wires), τ_{LB} is sub-femtosecond. This is a great deal faster than the period, $2\pi/\omega_\lambda$ of the environmental vibrations or polarizations that provide localization. In this limit, then, the mechanism should be coherent tunneling; the only vibronic coupling should be the very weak inelastic electron tunneling signal [64–68]. Conversely, when N is larger and E_G far smaller (π-type bridges), the value of τ_{LB} can approach $2\pi/\omega_\lambda$. Then the vibronic coupling to the λ mode can be significant (for large M_λ^{ij}, and activated polaron-type hopping should be seen). These timescale arguments are useful for understanding the length and energy dependence of transport mechanisms.

3 Factors that Determine the Charge Transfer Mechanism

3.1 Electronic Coupling

Both incoherent and superexchange mechanisms for electron transport between metallic or molecular donors and acceptors depend on the electronic coupling between the system's components. For the superexchange mechanism, the relevant quantity, as mentioned previously, is the donor–acceptor superexchange coupling, t_{DA}. If the bridge is energetically accessible and the incoherent mechanism dominates, the coupling between hopping sites is rate-determining.

3.1.1 Calculation of Electronic Coupling

The calculation of direct and indirect (superexchange) coupling has been a theoretical challenge. It has been formulated through generalized Mulliken-Hush (GMH) theory [69, 70] in terms of parameters that can be either derived from optical spectra in cases where the CT state is optically-accessible, or calculated using any method that yields excited state energies and dipole moments. However, for long-distance ET systems, the CT states from which t_{DA} can be obtained are not CT states at all, but rather real radical ion pairs (RPs) that can be several electron-volts above the neutral ground state and have vanishing transition moments. As a result, they display no emission upon charge recombination and acceptable determination of GMH parameters becomes very difficult.

There are several alternative approaches. Kumar and coworkers [71] used Marcus and Jortner's single-mode, semiclassical model for nonadiabatic ET to extract the electronic coupling for individual ET processes by calculating the free energy for reaction and the internal and external reorganization parameters.

Stuchebrukov et al. [72] calculated electronic superexchange matrix elements in several ruthenium-modified proteins, where the ruthenium, attached to the surface of the protein, donates an electron to a heme inside the protein 13–20 Å away. By modeling the structure of the donor and acceptor through ligand field theory using available spectroscopic data, they were able to use symmetry considerations to determine that the ET takes place mostly through a system of overlapping π orbitals. They began with the expression for t_{DA} in terms of the Green's function of the bridge (the intervening protein medium) in the nonorthogonal extended Huckel basis:

$$t_{DA}(E) = \sum_{ij} \langle A| V |i\rangle \left(ES - H^{B}\right)^{-1}_{ij} \langle j| V |D\rangle \tag{6}$$

where V is the strength of the coupling to the bridge, $S = \langle i|j\rangle$ is the overlap matrix and H^{B} is the Hamiltonian of the bridge in the extended Huckel basis set $|i\rangle$. However, instead of inverting the matrix $ES - H^{B}$, as would be done in Green's function methods, they rewrote t_{DA} in terms of transition amplitudes, T_i, the probabilities of "superexchange transitions" between the donor orbitals and the orbitals $|i\rangle$ of the bridge:

$$t_{DA}(E) = \sum_{i} \langle A| V |i\rangle \, T_i \tag{7}$$

To obtain the amplitudes T_i, one only has to solve a system of linear equations,

$$\langle i|\, V\,|D\rangle = \sum_j (ES - H^{\mathrm{B}})_{ij} T_j\,, \tag{8}$$

for which they recommend the conjugate gradient method [73, 74]. They get very reasonable agreement with effective superexchange coupling matrix elements derived from measured rates of ET with this method. An extension of the above treatment also allows one to distinguish hole pathway contributions from electron pathway contributions to the overall coupling [75].

The ample experimental data (see Sect. 5.3) on charge migration in DNA provides an excellent opportunity to test methodology for calculations of intrastrand and interstrand base-base couplings. Voityuk et al. [76] calculated nearest-neighbor intrastrand and interstrand couplings, which may then be used as inputs in the McConnell formulation for superexchange. These direct couplings were taken as half the splitting between ground and CT excited states of a base–base pair, which may be estimated within the Hartree–Fock–Koopmans' theorem formalism as the splitting between one-electron energies of the highest occupied molecular orbital (HOMO) and HOMO-1 orbitals of the neutral pair [77]. The distance dependence of coupling strength was exponential, and the results were found to be almost independent of the chosen basis set and to agree well with experimentally derived couplings.

Troisi and Orlandi [78] also calculated intrastrand and interstrand adjacent base couplings at the Hartree–Fock level as the off-diagonal matrix elements of the Kohn–Fock operator within a density functional theory calculation. They found that couplings between different pairs of bases may differ by up to an order of magnitude and that the geometry of the base pair can have a dramatic effect on the coupling magnitude.

3.1.2
Estimation of Donor–Acceptor Coupling via Magnetic Exchange Measurements

In certain molecular systems, there is a sensitive and quantitative way to measure t_{DA}. When the charge transport process originates from a state in which the redox centers are also paramagnetic, e.g., charge recombination from a RP, the electronic coupling that dictates CT from the RP to energetically proximate electronic states is also that which facilitates the magnetic exchange interaction between the unpaired spins of the RP [9, 10, 79–81]. Exchange interactions between magnetic moments of unpaired spins have been investigated to explain magnetic behavior in transition metal-insulator systems for decades [10, 11, 79, 82], and have recently sparked the interest of those attempting to build organic molecular magnets [83–86]. Sensitive measurement of exchange interactions within long-range RPs in electron transferring systems, most notably photosynthetic systems [87,88], using electron paramagnetic resonance [89] and optically detected magnetic resonance (magnetic field effects) [27,90,91] has sparked new interest in acquiring a fundamental understanding of indirect exchange mechanisms. The contribution of exchange coupling to the mechanism and efficiency of long-range charge and energy transfer processes [92], and the relationship between coupling and molecular structure [93–96], constitutes a central issue in constructing systems for use in molecular electronics.

The singlet and triplet RP states are either stabilized or destabilized through one- and two-step virtual CTs via coupling of the orbitals on the paramagnetic centers to the bridge orbitals and to each other. The total perturbation to each RP state, ΔE_S or ΔE_T for the singlet and triplet, respectively, is a sum of pairwise interactions between the RP state and the state to which it couples via CT . Anderson [11] used a perturbation approach to describe kinetic exchange between two magnetic centers separated by a nonmagnetic medium as the mixing between the ground state of the system and a CT excited state via a virtual excitation. Further work [97] led to the conclusion that the most important CT in the kinetic exchange mechanism is that between half-filled orbitals of the magnetic pair, giving a contribution of $2t_{ij}^2/\Delta E$ to the magnitude of exchange. Here t_{ij}^2 is the transfer integral [52] between orbitals on magnetic centers i and j and ΔE is the energy difference between the ground state and the CT state. The transfer integral t_{ij} is equivalent to McConnell's superexchange coupling, t_{DA}; therefore, under certain assumptions, the magnitude of the magnetic interaction is directly proportional to t_{DA}^2.

The magnetic interaction between the spins S_1 and S_2 for paramagnetic centers 1 and 2 is written in the form suggested by Heisenberg, Dirac, and Van Vleck [98]:

$$H_{EX} = -2J \cdot S_1 \cdot S_2 \,, \tag{9}$$

where J is positive if the spins are parallel and negative if they are antiparallel. For two spin 1/2 particles, the eigenvalues of H_{EX} for $S = S_1 + S_2$ give:

$$E(S) - E(S-1) = -2J \cdot S \tag{10}$$

$$E_{S\text{-}T} = E(1) - E(0) = -2J \tag{11}$$

The singlet-triplet (S-T) splitting, $E_{S\text{-}T}$, within the RP is therefore given by the phenomenological parameter, $2J$, the magnitude of the indirect exchange interaction.

The quantity $2J$ can be measured directly using magnetic field effects (MFEs) on the yield of triplet states resulting from RP recombination. The MFEs are due to the radical pair intersystem crossing (RP-ISC) mechanism, which is well-known to account for triplet production within photosynthetic reaction centers [99–103], and has been described in detail elsewhere [27,91,104–110]. The total spin Hamiltonian for a radical pair in solution, where dipole–dipole coupling and other anisotropic interactions average to zero, is given by [111]

$$H_{ST} = \beta B_0(g_1 S_1 + g_2 S_2) + \sum_i a_{1i} \boldsymbol{S} \cdot \boldsymbol{I}_i + \sum_k a_{2k} \boldsymbol{S} \cdot \boldsymbol{I}_k + H_{EX} \quad (12)$$

where β is the Bohr magneton, B_0 is the applied magnetic field, g_1 and g_2 are the electronic g-factors for each radical, $\boldsymbol{S}_1$ and $\boldsymbol{S}_2$ are electron spin operators for the two radicals within the radical pair, $\boldsymbol{I}_i$ and $\boldsymbol{I}_k$ are nuclear spin operators, a_{1i} and a_{2k} are the isotropic hyperfine coupling constants of nucleus i with radical 1 and nucleus k with radical 2.

Immediately following charge separation from a singlet excited state precursor, the correlated electron spins are in a singlet configuration. This pure state is, in general, not an eigenstate of H_{ST}, as the weakly coupled electron spins are free to precess independently around the resultant of their respective local fields (due mainly to electron–nuclear hyperfine interactions) and the external applied field. After times usually in the range of a nanosecond, this RP-ISC results in formation of a triplet spin configuration. When hyperfine and exchange interactions are isotropic and spin–spin coupling is weak, and if $B_0 = 0$, each of the three zero-field triplet states of the RP will be nearly degenerate with the singlet and will be populated with equal probability at room temperature. If the spin–spin exchange interaction within the RP is non-zero, the triplet manifold is not initially degenerate with the singlet, but rather separated from the singlet by an energy $2J$.

Application of a magnetic field results in Zeeman splitting of the triplet sublevels, which at high fields can be described by the T_0 and $T_{\pm 1}$ states (see Fig. 2). In the high field limit, population of the RP triplet state occurs exclusively by S-T_0 mixing, while T_{-1} and T_{+1} remain unpopulated. When the Zeeman energy from the applied field equals the S-T splitting, either the low-energy triplet state, T_{-1}, or the high-energy triplet state, T_{+1}, depending on the sign of $2J$, crosses the singlet, and the RP-ISC rate is maximized, which produces a resonance in the triplet yield at magnetic field value B_{2J}. An increase in the rate of triplet formation at resonance implies that the RP decay rate also increases. One can therefore monitor the RP population as a function of the applied magnetic field and obtain a plot with a minimum at B_{2J} to obtain $2J$ as well. The $2J$ resonance gives the zero-field splitting of the singlet and triplet RP levels due to magnetic superexchange coupling.

A clear example of the exponential decay of superexchange coupling with r_{DA},

$$2J = 2J_0 \exp(-\alpha r_{DA}) \quad (13)$$

can be found in the case of the series of donor–bridge–acceptor (DBA) systems, phenothiazine-p-oligophenylene- perylene diimide (PTZ-ph$_n$-PDI, $n = 2, 3, 4, 5$) [27]. In Fig. 3, ln $2J$ is plotted versus r_{DA}, with fitting parameters $\alpha = 0.37$ Å^{-1} and $2J_0 = 8.6 \times 10^4$ mT. Because, as previously mentioned,

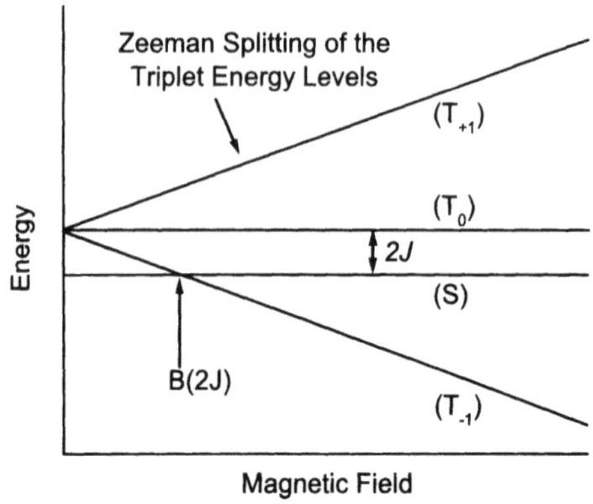

Fig. 2 Schematic of radical ion pair energy levels as a function of magnetic field

the superexchange coupling t_{DA} is related to $2J$ through the approximate expression $2J \sim 2t_{DA}^2/\Delta E$, and there is very little distance dependence of ΔE, t_{DA} decays with $\beta = \alpha/2 = 0.19\,\text{Å}^{-1}$ for this system. This measurement provides quantitative evidence that the magnetic superexchange coupling, and therefore the probability of a coherent ET mechanism, decreases approximately exponentially as the bridge lengthens. Concurrently, the near resonant interaction of the radical pairs PTZ^+-ph_n-PDI^- and PTZ-ph_n^+-PDI^-, when $n = 4$ and 5, makes incoherent hopping a viable CT mechanism. These two effects combined ensure the dominance of wire-like charge transport (incoherent) for n=4 and 5, as will be discussed in Sect. 5.2. These results allowed for the first observation of the relative contributions of both the coherent superexchange and incoherent hopping mechanisms to the overall charge transport process in a conjugated bridge molecule.

Studies concerning couplings between paramagnetic metal centers [112–114] and biradicals [115–117] have shown that multiple CT pathways (σ–σ, σ–π, π–π) may be contributing to the S-T splitting, but previous work in calculating $2J$ has shown [95] that assigning different distance dependencies to different CT pathways in hopes of distinguishing the respective orbitals involved does not improve the fit to experimental data over using a single exponential. Although it is not straightforward to deconvolute the separate contributions of various CT pathways to the total S-T splitting, both $2J$ and t_{DA} are inherently a sum of many ET processes. The exponential decrease of $2J$ with r_{DA} parallels the approximately exponential decay of the coherent contribution to the various ET processes.

The $2J$values are measurements of effective coupling: average values, taken over the conformational space sampled by the molecule [118]. The magnetic field effect experiment can also be used to monitor this effective superexchange coupling as a function of temperature. The magnetic field effect on yield of triplet recombination product within the donor–chromophore–acceptor system, , where p-methoxyaniline– 4-(N-piperidinyl)naphthalene-

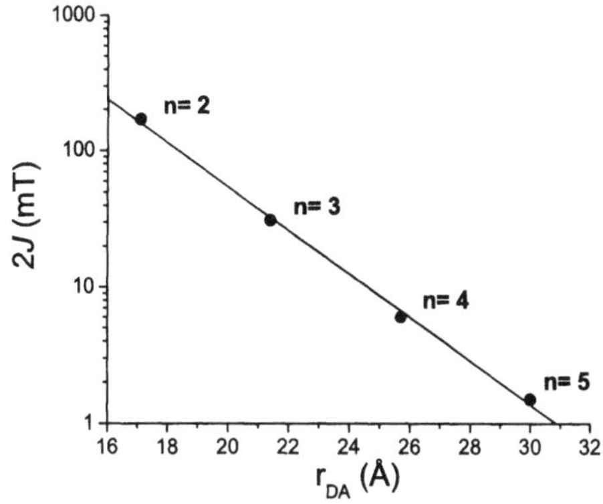

Fig. 3 Logarithmic plot of the spin-spin exchange interaction, $2J$ vs. r_{DA} for the phenothiazine–p-oligophenylene–perylene diimide (PTZ–ph$_n$–PDI) species [176]. The best fit line through the data points for $n = 2 - 5$ gives $R^2 = 0.99$ and a slope of -0.37 Å^{-1}

1,8-dicarboximide- naphthalene-1,8:4,5-bis(dicarboximide) (MeOAn-6ANI-NI) was measured from 140 to 300 K in toluene [119]. Two distinct groups of RP conformations are identified by the approximately fourfold difference in magnitude of the spin–spin exchange interaction, $2J$, at 140 K, which directly reflects the difference in their electronic coupling. The equilibrium shifts almost entirely to the more strongly coupled conformation as the temperature is increased.

Paddon-Row and Shephard [118] used time-dependent density functional theory to calculate S-T gaps for charge-separated states of a series of DBA molecules, with a dimethoxynaphthalene donor, a dicyanovinyl acceptor and a norbornyl bridge. They used a variety of functionals (B3P86, B3LYP, B3PW91, BPW91, and BLYP) [120] and a 6–311 G(d) basis set and calculated an exponential decay of $2J$ with $\alpha = 0.91$ per bond, in good agreement with experimental results. Nelsen and coworkers [121] confirmed the asserted relationship, $2J = 2t_{\mathrm{DA}}^2/\Delta E$, between the S-T splitting and the electronic coupling for CT within a symmetrical, strongly coupled bishydrazine. The electronic coupling and the energy of the CT state were calculated via Mulliken-Hush theory, and $2J$ determined by electron spin resonance.

3.2 Energy Matching

A key to assessing the transport properties of a system is the determination of the charge injection energy, given by the energies of the frontier orbitals of the molecular bridge relative to the Fermi energy of the metallic contacts or the molecular levels of the donor and acceptor. This is a slightly misleading but very common description. What is really relevant is the state energies of the donor (or the Fermi level) and of the bridge cation and bridge anion. The frontier orbital energies are often a reasonable approximation to these multi-electron values. In particular, electron transport across metal–molecule–metal junctions depends strongly on the position of the Fermi level of the metal electrodes relative to the lowest unoccupied molecular orbital (LUMO) and the HOMO of the molecular bridge. When the difference in energy between the LUMO and the Fermi level is large, electron transport occurs by superexchange tunneling, i.e., tunneling mediated by interaction between donor and acceptor and unoccupied orbitals of the organic bridge that separates them. If the Fermi level approaches the energy of the orbitals of the molecular bridge, resonant ET may take place (either hopping or resonant tunneling) and the conduction of electrons will occur through the molecular orbitals.

A clear example of the importance of energy matching in determining transport mechanism can be found in the work of Fan et al. [122], who used shear-force-based scanning probe microscopy (SPM) combined with current–voltage (I–V) and current–distance (I–d) measurements to investigate interfacial ET across self-assembled monolayers (SAMs) of hexadecanethiol and nitro-substituted compounds, including oligo(phenylene ethynylene) (OPE), on gold. The hexadecanethiol SAMs showed an exponential increase in current with decreasing distance with a relatively large β value, ranging from 1.3 to 1.4 Å^{-1}. This dependence was nearly independent of the tip bias in the low bias regime and the current increased exponentially with bias at high bias voltage. All of these observations suggest that transport through the hexadecanethiol SAMs proceeds through a superexchange mechanism due to the mismatch between molecular orbital energies and the Fermi level of gold.

However, in the case of the nitro-substituted OPE SAMs, where the molecular orbitals are in close energetic proximity to the Fermi level, a different observation was made. In this case, the current was more weakly distance-dependent and the β value low and strongly dependent on the molecular structure. Reversible peak-shaped I–V characteristics, often described as negative differential resistance, were obtained for most of the nitro-substituted OPE SAMs studied here, indicating that part of the conduction mechanism of these junctions involves resonant tunneling. Although the current decreased with increasing distance, d, between tip and substrate, the β-values for the OPE molecules were low (0.15 Å^{-1} at a – 3.0-V tip bias) and depended strongly on the tip bias, because the molecular orbitals are affected by interaction with the contacts and by the applied voltage.

Photoelectron-spectroscopy-related methods provide a direct means of addressing the issue of energy matching. One-photon photoelectron spectroscopy provides access to the energies and characteristics of the occupied electronic levels while two-photon photoelectron spectroscopy allows investigation of unoccupied levels. Zangmeister et al. [123] used one- and two-photon ultraviolet photoelectron spectroscopies to determine the electronic structure around the Fermi level for SAMs of 4,4-(ethynylphenyl)-1-

benzenethiol (an OPE) on gold. One-photon ultraviolet photoelectron spectroscopy yielded a hole injection barrier of 1.9 eV and an electron injection barrier of 3.2 eV. The shift toward the HOMO from mid-gap is due to CT interactions between sulfur and the gold surface [124].

4
Methods of Calculating Conductance

Formulations for estimating the nonadiabatic ET rate for a variety of energy and coupling configurations, first developed by Marcus, Hush, and Jortner, are well-documented [7, 125]. The nonadiabatic ET through a molecule from donor, D, to acceptor, A, is

$$k_{\mathrm{DA}} = \frac{2\pi}{\hbar} |V_{\mathrm{DA}}|^2 \, FCWD \tag{14}$$

where V_{DA} is the donor–acceptor superexchange coupling and $FCWD$ is the Franck–Condon weighted density of states. In the classical limit [126–129],

$$FCWD(\Delta E) = \frac{\mathrm{e}^{-(\lambda+\Delta E)^2/4\lambda k_{\mathrm{B}} T}}{\sqrt{4\pi\lambda k_{\mathrm{B}} T}} \tag{15}$$

where λ and ΔE are the reorganization energy and free energy change, respectively, of the CT reaction.

More recent developments have been in the theory used to calculate conductance across a molecule, the rate of CT in the case where donor and acceptor are a continuous density of states across which a time-dependent voltage is applied. The coherent, elastic conductance, g, is most often approached through the Landauer formula [130–133]

$$g = \frac{2e^2}{h} \sum_i T_i \, , \tag{16}$$

where T is the scattering matrix, whose elements give the probability of the electron's scattering via channel i. Of course, if the transmission probability is unity for every channel, then the conductance is an integral multiple of the quantum of conductance, $2e^2/h = (12.8\ \mathrm{k}\Omega)^{-1}$. The scattering matrix can be expanded in the Green's function of the molecule, G,

$$T = \mathrm{Tr}\left\{\Gamma G^{\mathrm{A}} \Gamma G^{\mathrm{R}}\right\}; \quad G^{\mathrm{A}}(E) = [G^{\mathrm{R}}(E)]^{+} = (E - H_{\mathrm{m}} + i\varepsilon)^{-1} \tag{17}$$

where H_{m} is the molecular Hamiltonian, and Γ is the spectral density coupling the molecular terminus to the electrode. Conductance measurements are far newer than rate constant measurements in donor–bridge–acceptor (DBA) species. Moreover, while many test beds exist for measuring conductance in molecular transport junctions, the interpretation remains challenging, mostly because contact effects can totally dominate (the contact-dominated spectral density Γ term in Eq. 17 varies more from measurement to measurement than does the molecular contribution G). Both experiments and theory agree that in some limits (notably oligoalkane bridges [134–137]), the transport is simple tunneling, but even for that case disorder effects cause lack of agreement from experiment to experiment. With smaller energy gaps [e.g., oligo(phenylenevinylene) (OPV) wires [138]), the Landauer formula fails because charges localize on the wire and move by a polaron-type mechanism [139, 140] that does not conform to the elastic-scattering limit assumed by Landauer. A far more general nonequilibrium Green's function technique has been useful for characterizing vibronic effects in junction transport [130, 132, 133, 141–144].

Nitzan [2, 145] discussed the relationship between conduction through a metal–insulator–metal (MIM) junction and the nonadiabatic ET rate through the molecule serving as the insulator. In the case where the DBA system with identical bridge units, 1...N with nearest-neighbor couplings V_B, V_{D1}, and V_{NA}, is the molecular conductor between two metallic leads, the Hamiltonian for the system can be written:

$$H = H_0 + V \tag{18}$$

where H_0 describes the uncoupled electrodes and the molecule and V is the coupling between them. Then, the scattering matrix can be written, in the absence of direct coupling between metallic leads:

$$T(E) = VG(E)V \tag{19}$$

where $G(E) = (E - H - i\varepsilon)^{-1}$ is the retarded Green's function for the system. The ET rate through the molecular linker is analogous to Eq. 14,

$$k_{DA} = \frac{2\pi}{\hbar} \left| V_{D1} V_{NA} \right|^2 \left| G_{1N}(E_D) \right|^2 FCWD \tag{20}$$

where G_{1N} is an element of a submatrix of G corresponding to the bridge subspace. When $|V_B| \ll |E_B - E|$ (the weak coupling limit), then

$$G_{1N}(E) = \frac{\left| V_B \right|^{N-1}}{(E_B - E)^N} \tag{21}$$

If the donor and acceptor are strongly coupled to the left (L) and right (R) leads, then the conduction through the junction is

$$g(E) = \frac{e^2}{\pi\hbar} \left| G_{DA}(E) \right|^2 \Gamma_D^{(L)}(E) \Gamma_A^{(R)}(E) \tag{22}$$

$$G_{DA}(E) = \frac{V_{D1} V_{NA}}{(E - E_D - \Sigma_D(E))(E - E_A - \Sigma_A(E))} G_{1N}(E)$$

where Γ is defined above and Σ_D and Σ_A are the components of the total self energy of the molecule associated with the perturbation of the donor due to the left contact and the acceptor due to the right contact, respectively. If we assume that the structure of the molecule is not altered by the contacts, then the conductance for injection at the Fermi level of the metal, E_F, may be expressed as:

$$g = g(E_F) = \frac{16e^2}{\pi\hbar} \frac{|V_{D1} V_{NA}|^2}{\Gamma_D^{(L)}(E_F)\Gamma_A^{(R)}(E_F)} |G_{1N}(E_F)|^2 \; ; \quad E_F = E_D = E_A \tag{23}$$

Then,

$$g = \frac{e^2}{\pi\hbar} \frac{k_{DA}}{FCWD} \frac{8\hbar}{\pi \Gamma_D^{(L)} \Gamma_A^{(R)}} \tag{24}$$

where k_{DA} is given by Eq. 20. For reasonable values of $FCWD$ $(0.02\,\text{eV})^{-1}$, and $\Gamma_D = \Gamma_A$ (0.5 eV), $g \approx [10^{-17}\ k_{DA}(\text{s}^{-1})]\Omega^{-1}$ such that k_{DA} must exceed $\sim 10^6\ \text{s}^{-1}$ in order to observe measureable current (with a detector sensitive to pico-amperes) at – 0.1 V voltage.

5 Examples of Molecular Wire Systems

Charge and energy transport through a variety of potential molecular wire systems, e.g., benzenes [146, 147], modified proteins and peptides [148, 149], *p*-phenylacetylenes [32, 150, 151], and porphyrin arrays [152–155], have been investigated. The methods of investigation are just as varied. SAMs [156, 157] have been the basic building block for the majority of organic devices. Various types of MIM junctions [158] including single layers of molecules between aluminum and titanium/aluminum [159, 160], gold and other metal contacts [2, 161], mechanically controllable break junctions [146], silicon adlayers [162], electrochemical break junctions [163], and SAMs sandwiched between two mercury electrodes [164–166] have all facilitated conduction measurements and, at the same time, provided an extensive set of transport environments leading to a wide range of results. In this section, we will touch on just a few example systems for energy transfer and for conductance.

5.1 Oligo(phenylenevinylene)

OPV is considered an excellent choice for efficient charge transport because it is composed of coplanar phenyl groups linked by vinylene bonds around which the phenyl groups have a higher barrier of rotation than in, for example, the OPE system [167]. This conformational rigidity preserves the conjugation of the π-system. Room temperature current-voltage (I–V) measurements of gold nanowire–SAM–gold nanowire junctions [168, 169] showed that for a particular pair of metal contacts the conductance of OPV molecular wire junctions is approximately one order of magnitude larger than the OPE molecular wire junctions and three orders of magnitude larger than the saturated dodecane junctions.

Dudek and coworkers [170] synthesized ferrocene OPV methyl thiols and characterized their SAMs on a gold electrode. Through a laser-induced temperature jump technique, they found that CT to the tethered ferrocene through even the longest oligomer (35 Å) proceeded at a rate greater than $10^5\ s^{-1}$, and interpreted the mechanism to be fast electron-tunneling, as the energy levels of the bridge relative to the gold ruled out a hopping mechanism [34]. They note that extrapolation of the distance-dependent rate constants reported for ferrocene ester-linked alkanethiols to the same length gives a rate constant of $4 \times 10^{-5}\ s^{-1}$, 10 orders of magnitude slower than the OPV systems. They suggest that, for bridges 28 Å or shorter, it is structural reorganization of the redox species with respect to the monolayer–water interface, not electronic coupling, that is rate limiting.

A molecule may show very different transport behaviors depending on the environment. Sikes et al. [34] have studied transport through OPV and found that efficient tunneling occurs between a gold surface and an acceptor through OPV over 28 Å, during which no electrons are donated or removed from the bridge. In contrast, Davis et al. [3, 171] studied a series of five donor–bridge–acceptor (DBA) molecules in which the donor was tetracene, the acceptor was pyromellitimide, and the bridge molecules were OPVs of increasing length. For short bridges (OPV_1 and OPV_2) superexchange dominated charge separation, which was strongly distance-dependent. But for longer bridges, bridge-assisted hopping dynamics dominated, resulting in relatively soft distance dependence for ET and a rate 3–5 times higher than in the shorter bridges.

The temperature dependence of the charge separation rates in all five molecules in the Davis studies did not appear to obey the predictions of semi-classical ET theory. The hypothesis that ET was "gated" by torsional motion between the tetracene donor and the first bridge phenyl ring was put forth and supported by the fact that the measured activation energies for charge separation for all five molecules equaled the frequency of a known vibrational mode in 5-phenyltetracene. Additionally, in the molecule containing a *trans*-stilbene bridge, a competition exists between the tetracene–phenyl torsional motion and one that occurs between the vinyl group and the phenyls linked to it, resulting in complex temperature-dependence of charge separation that exhibited both activated and negatively activated regimes. The occurrence of higher rates in the molecules with longer bridges was attributed to increased planarity between donor and acceptor.

Subsequent theoretical work [172], in which density functional theory was used to study the electronic structure of OPV and quantum dynamics calculations were used to study the ET, supported the switch in mechanism from tunneling through the bridge to utilization of the bridge as a real intermediate. With a model that included no localization of charge on the shorter bridges, they obtained rate constants that agreed well with those experimentally observed by Davis. However, they found that it is not only torsional modes of the bridge plane or side groups, but also C – C stretch vibrations that contribute to the higher rates in the longer bridges.

DBA systems with extended tetrathiafulvalene donors, OPV bridges, and fullerene acceptors have been constructed [173, 174]. Superexchange-mediated hole transfer occurs over distances of up to 50 Å with a very small attenuation factor $\beta = 0.01 \pm 0.005$ Å^{-1}. The authors attribute this efficient CT to energy matching between the HOMOs of the excited fullerenes and those of the OPV bridges and to strong electronic coupling (they estimate 5.5 cm^{-1}) between the donor and acceptor moieties, mostly due to donor–bridge orbital overlap.

5.2 Oligophenylene

Schlicke et al. [175] synthesized rod-like compounds in which two metal-bipyridyl complexes were linked with oligophenylene. The metal-to-metal distance was 4.2 nm for the longest spacer (seven phenylene units). The rate of the Dexter-type energy-transfer process from the $[Ru(bpy)_3]^{2+}$ to the $[Os(bpy)_3]^{2+}$ unit was found to be essentially temperature-independent and decreased exponentially with an attenuation coefficient of 0.32 Å^{-1} (1.5 per interposed phenylene unit).

It has been observed [176] that the aforementioned oligomeric *p*-phenylene bridge (ph_n) acts as a molecular wire for the charge recombination reaction of phenothiazine-*p*-phenylene-perylenediimide, PTZ^+-ph_n-PDI^-, when $n \geq 4$. The rate constants for charge recombination were obtained from the decay rate of the PDI anion. Charge recombination within the DBA compounds where $n = 1$, 2, 3 was strongly exponential, and the plot of log (k_{CR}) vs r_{DA} yielded $\beta = 0.67$ Å^{-1}, Fig. 4. The rate constants for charge recombination in the compounds where $n = 4$ and 5, however, did not lie along the linear curve established by $n = 1$, 2, 3, and, in fact, actually *increased* with increasing bridge length. For charge recombination within these compounds, a switch in mechanism from superexchange to thermally activated hopping as the bridge is lengthened is postulated.

A change in charge recombination mechanism from superexchange to hopping relies on efficient charge injection into the ph_n bridge such that the $PTZ^{+\cdot}$-$ph_n^{-\cdot}$-PDI and/or PTZ-$ph_n^{+\cdot}$-$PDI^{-\cdot}$ states are real intermediates. The energies of PTZ^+-ph_n^--PDI for $n = 1$–5 all are ≥ 3.0 eV, so that electron injection onto the bridge during the charge recombination reaction cannot occur. On the other hand, as the bridge lengthens, the calculated energy of PTZ-$ph_n^{+\cdot}$-$PDI^{-\cdot}$ decreases significantly due to the increased conjugation length, as can be seen in the decrease in oligophenylene band gap energy

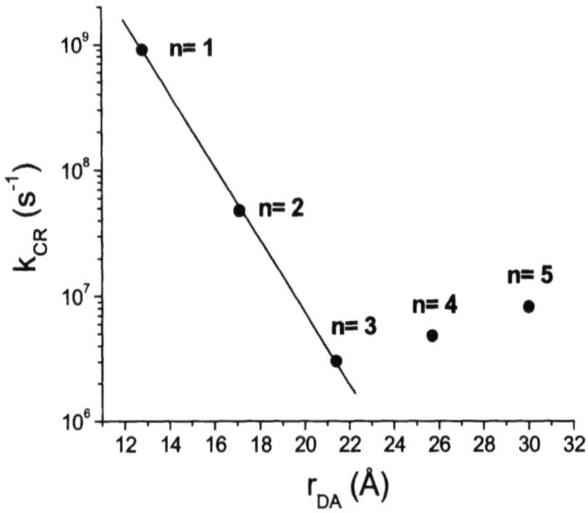

Fig. 4 Logarithmic plot of the charge recombination rate constant, k_{CR} vs. donor–acceptor distance, r_{DA} for the PTZ–ph_n–PDI molecular wire system [176]. The best fit line through the data points for $n = 1, 2, 3$ gives $R^2 = 0.99$, $\beta = 0.67$ Å^{-1} and $k_0 = 5 \times 10^{12}$ s^{-1}

from 7.3 eV for $n = 1$ to 3.3 eV for $n = 5$, calculated with time-dependent DFT (B3LYP hybrid functional, 6–31 G** basis set) using the Q-Chem 2.1 software package [177]. The accompanying increased ease of oxidation of the oligo-*p*-phenylene, coupled with the increase in the energy of $PTZ^{+\cdot}$-Ph_n-$PDI^{-\cdot}$ due largely to Coulomb destabilization, Fig. 5, leads to a near resonance of PTZ-$ph_n^{+\cdot}$-$PDI^{-\cdot}$ with $PTZ^{+\cdot}$-ph_n-$PDI^{-\cdot}$ for $n = 4$ and 5. The slight increase in charge recombination rate observed for $n = 4$ and $n = 5$, relative to $n = 3$ may be due to the combination of the increasing electronic interaction between $PTZ^{+\cdot}$-ph_n-$PDI^{-\cdot}$ and PTZ-$ph_n^{+\cdot}$-$PDI^{-\cdot}$ as the energy gap between them be-

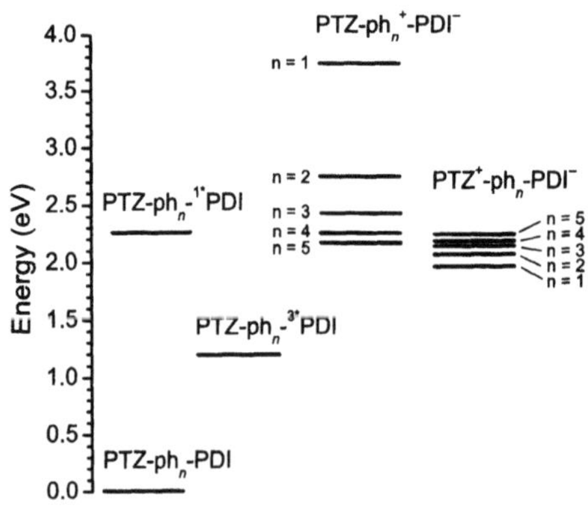

Fig. 5 Energy levels for the electronic states relevant to the charge recombination pathways for the PTZ–ph_n–PDI, $n = 1$–5

comes smaller and the decreasing internal reorganization energy associated with putting a charge on a longer bridge.

5.3 DNA

It has been proposed that β should be ≤ 0.2 Å^{-1} for DNA to be considered a molecular wire [178]. However, many (e.g. [179]) have proposed that superexchange and charge hopping are competitive in DNA, and the mechanism depends on the energy gap between donor and bridging states, as well as the sequence, and therefore varies with experiment.

Berlin and coworkers [180] asserted that hole-hopping between G bases is the dominant mechanism of CT within DNA and that, for long, regular sequences of AT and GC base pairs, the CT rate, as measured by hydrolytic cleavage [181], will fall off with a shallow exponential distance dependence ($\beta \sim 0.1$ Å^{-1}). This behavior is indicative of the hydrolytic decay as it moves through the strand, not of interbase electronic coupling. The same group [14, 182] studied the effects of injection barrier and base sequence on hole migration within DNA, and included dynamic disorder by treating the base pairs as a series of harmonic oscillators coupled to the injected charge and numerically propagating the wavefunction of the charge, see Fig. 6. For high injection barriers (0.55 eV), they found that single-step tunneling dominated transport and that this mechanism was exponentially distance-dependent with a large fall-off parameter ($\beta \sim 0.85$ Å^{-1}). When the injection barrier was lowered, however, β dropped significantly to a value of 0.09 Å^{-1} for the limiting case of a 0 eV barrier. Additionally, they found that interrupting a relatively short AT base sequence with a GC pair increased the CT rate by almost two orders of magnitude and weakened the distance dependence by shortening the average length of the tunneling steps that comprise the hopping transport. In longer chains, however, thermal activation may dominate tunneling as the mechanism of G $\rightarrow$G transport such that the average rate is no longer sequence dependent [13]. Bixon and Jortner have discussed general thermal hopping models [183], and given specific comparisons to distance-dependent measurements, while Rösch's group [184] has used electronic structure methods to predict both the energetics of hole formation and the interbase tunneling matrix elements.

Predictably, a range of values for β for charge separation and charge recombination have been observed, $\beta = 1.5$ Å^{-1} [185], $\beta = 1.0$ Å^{-1} [186], $\beta = 0.9$ Å^{-1} [187], $\beta = 0.7$ Å^{-1} [188], reflecting the influence of the charge injection process and the DNA sequence chosen. Lewis et al. [189] covalently bound stilbenedicarboxamide to small, synthetic DNA hairpins and found β to be 0.7 Å^{-1} for the forward ET and 0.9 Å^{-1} for the reverse step. It was suggested that these relatively small βs might be due to the similarity in

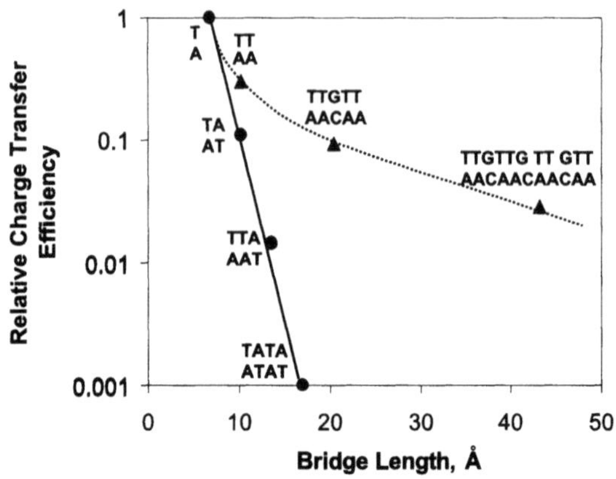

Fig. 6 Hole transport rates in DNA, as measured using hydrolytic cleavage. Note that the bridges containing only A and T bases behave like simple tunnel barriers, while GA pairs cause localization and hopping transport. The *solid triangles* are experimental data from [216]; the *solid circles* are experimental data from [217]. The *dotted line* is theory for superexchange, *solid line* is theory for hopping [182]. The sequences over which the charge transfer efficiency was measured or calculated (between G^+ and GGG) are listed next to the data points

energy of the HOMO in stilbene to that in DNA. Wan et al. [190] incorporated 2-aminopurine, an isomer of adenine, into well-characterized DNA assemblies, and a similar distance dependence to that of Lewis was observed, $\beta = 0.6\,\text{Å}^{-1}$.

Okamoto and coworkers [191] incorporated designer nucleobases, benzodeazaadenine derivatives, which have lower oxidation potentials and wider stacking areas than natural bases, into a DNA strand. They observed efficient hole transport between two GGG sites that are 76 Å apart without any detectable nucleobase decomposition and concluded that highly ordered π-stacking, in addition to low oxidation potentials, and minimal oxidative degradation are necessary for DNA to act as a wire.

Recent work by Tao and collaborators [192] has demonstrated both strong and weak distance dependence of conductance in DNA double strands in electrochemical break junctions. The overall behavior observed (as is generally true of both CT and conductance measurements [193]) is consistent with a combination of local tunneling of holes in GC regions, with far less efficient tunneling, or activated hopping, over AT regions [183, 194].

5.4 Oligothiophene

Synthetic work on oligothiophenes with more than ten thiophene units was begun in the early 1990s, when Wynberg et al. in 1991 reported the synthesis of the region-irregular decyl-substituted 11-mer [195]. Since then, oligothiophenes have been made with sufficient length to span two nanoelectrodes made by lithographic techniques; Otsubo and coworkers synthesized a 72-mer in 2001 [196]. It was observed that, even for the longer 15-mer [197] and 16-mer [198], the ground state absorption energy continued to red-shift. An extrapolation of the absorption energy/chain length relationship suggested that the effective conjugation length is 20 thiophene units in the neutral state and around 30 units in the oxidized state [199].

The high degree of conjugation in these molecules has made them attractive candidates for bridges in a number of multi-component systems such as anthryloligothienylporphyrins [200, 201] and anthryloligothienylfullerenes [202, 203]. Sato et al. [204] built hexyl-sexithiophene and methoxy-terthiophene derivatives with two terminal ferrocenyl groups as model compounds for molecular wires. In the hexyl-sexithiophene derivative, the resultant oxidized state spread over both the ferrocene and the sexithiophene moiety. In the methoxy-terthiophene derivative, the oxidized species spread over the entire molecule containing the terthiophene moiety and the other ferrocene moiety terminal, indicating the CT between the terminals via the wire.

Otsubo and coworkers [205] examined porphyrin-oligothiophene-fullerene triads, in which rates of intramolecular ET from the porphyrin to the fullerene were estimated for different oligothiophene lengths from the degree of quenching of the porphyrin fluorescence. The ET took place on the nanosecond timescale: $5.7 \times 10^9\ s^{-1}$ for $r_{DA} = 1.4$ nm, $6.2 \times 10^8\ s^{-1}$ for $r_{DA} = 3.0$ nm, $2.0 \times 10^8\ s^{-1}$ for $r_{DA} = 4.6$ nm. A plot of the natural log of the rate constant versus r_{DA} is linear with a decay parameter of ~ 0.11 Å^{-1}.

Nakamura and coworkers [206] studied photo-induced charge separation (CS) and recombination (CR) processes of porphyrin (H2P)–oligothiophene (nT)–fullerene (C60) linked triads, where $n = 4$, 8, and 12, by time-resolved fluorescence and absorption spectroscopic methods. After the excitation of the H2P moiety in benzonitrile and *o*-dichlorobenzene, $H2P^{+}$-nT-C_{60}^{-} was produced and the subsequent hole shift resulted in the energetically stable H2P-nT^{+}-C_{60}^{-} state. The initial charge separation rate decreased with bridge length with a small, solvent-dependent exponential attenuation factor (0.03 Å^{-1} in benzonitrile and 0.11 Å^{-1} in *o*-dichlorobenzene).

In order to model the effects of incorporating oligothiophene bridges into solid-state devices via thiol linkages, Hicke, et al. [207] studied several oligothiophenes with 2-mesitylthio (MesS) substituents. MesS groups are strong electron-donating substituents and have been shown to strongly perturb the electronic structure of oligothiophenes by affecting their redox properties and constraining their geometry. These groups may act as hole traps or, alternatively, promote hole-hopping onto the polymer bridge. Due to conjugation between the mesitylthiosulfur lone pair and the oligothiophene π system, the ground state solution electronic spectra of all oligomers displayed a red-shifted lowest-energy transition maximum (λ_{max}) relative to oligothiophenes lacking the MesS group. Electrochemical measurements indicated that the MesS group significantly lowers the first and second oxidation potentials of the oligomers, and, by causing charge density in the radical cations to concentrate at the chain ends, lowers the Coulombic barrier to introduction of a second positive charge. As a result, the difference between first and second oxidation potentials in the MesS-substituted oligomers was much lower than known alkyl-substituted oligomers. Finally, these groups were shown to improve the stability of the bis(arylthio)oligothiophene radical cations such that they can be more robust charge carriers.

For a comprehensive review on oligo- and poly-thiophenes, see [208], and for more examples of surface-bound thiophenes, see [209, 210].

5.5 Photonic Wires

A photonic wire conducts excited-state energy rather than charge from donor to acceptor. Photonic wires may be operated from a distance, i.e., by the application of light without any physical contacts to the device. Lindsay and co-workers built the first molecular photonic wire based on conjugated porphyrin arrays [211] in 1994. More recently, Heilemann et al. [212] covalently attached five different chromophores to single-stranded DNA fragments of various lengths (60 or 20 bases) at well-defined positions and induced hybridization of the fragments. The chromophores were spaced 3.4 nm from each other for a total length of 13.6 nm. Single molecule spectroscopy in combination with spectrally resolved confocal fluorescence scanning microscopy on four detectors was used to get exclusive excitation of the first donor. Emitted photons were split into four spectrally separated collection channels. About 10% of all photonic wires show predominantly emission on the red channel, that is, unidirectional highly efficient ($\sim$ 90%) multistep energy transfer across 13.6 nm.

6 Remarks

Electron and energy transfer are two of the most significant processes in chemistry and in biology. By restricting the transfer to either within a given molecule (intramolecular electron or energy transfer kinetics) or across a given molecule (single molecule transport junction conductance), it is possible to characterize both the response coefficients (rate constant or conductance) and the mechanisms for such transfers. As we have seen, the dominant mechanisms extend from coherent tunneling (for large injection energy gaps, low temperatures, and short bridges) to incoherent polaron-type hopping (for small gaps, higher temperatures, and longer wires). The relative importance of these two can be understood on both energetic (the injection barrier versus thermal energy) and temporal (Landauer/Buttiker contact time versus vibrational period) bases.

A tight binding model linearly coupled to environmental boson excitations is given in Eq. 1: it describes very well the motion of both excitons and charges along molecular wires. Because excitons are (at least at low energies) just electron/hole pairs, it is not surprising that the mechanistic patterns of electron and energy transfer should be similar. It might also be expected that the matrix element for exciton transfer should be proportional to the square of the ET matrix element. As pointed out in Section 3, such is often the case. The similarities go even deeper: Hopfield [213] used ideas on Förster energy transfer to develop a formalism for calculation of ET rates, and Fleming and Scholes [214] used superexchange-type ideas from ET theory to derive approximate rate expressions for bridge-assisted long range energy wire behavior.

In classical chemical kinetics, one studies a thermal ensemble of systems. The rate constant for a given reaction can then be observed or calculated from Boltzmann distribution averages. For true molecular wire behavior, one is concerned with a single molecule observation. Then the transfer rate within a given donor/bridge/acceptor molecule, or the conductance of a single molecule transport junction, would be expected to show the characteristic fluctuations that are the hallmarks of single-molecule spectroscopy. Such fluctuations have indeed been observed in molecular conductance, both for a single system observed over time [215] and for multiple systems with the same components and roughly the same structure [163]. As in single-molecule spectroscopy, these fluctuations and blinkings almost certainly arise from different geometries of the bath particles or the atoms comprising the molecular wire or the terminal (donor, acceptor, or electrode) groups.

Many issues, besides those characteristic fluctuations, remain to be clarified for the understanding of molecular wire behavior. Ever greater challenges occur in trying to control, by synthesis, positioning, and excitation, the rate processes. This is crucial in the preparation of much of organic and molecular electronics, as well as photovoltaics, sensors and some medical intervention protocols. Molecular wires continue to pose major and tantalizing challenges to the science and technology communities.

Acknowledgements We are grateful to many members of our research groups, particularly Abe Nitzan, Vladimiro Mujica, Yuri Berlin, Randall Goldsmith, Louise Sinks, and Michael Ahrens for helping in our attempts to understand molecular wires. This work was funded by the MURI/DURINT program of the DOD, by the Chemistry Division of the NSF and ONR, by the NSF-NNI program through the Purdue-centered NCN, and the Division of Chemical Sciences, Office of Basic Energy Sciences, U.S. Department of Energy under grant no. DE-FG02-99ER-14999. E.A.W. would like to thank the Link Foundation and Northwestern University for fellowships.

References

1. Emberly EG, Kirczenow G (1998) Phys Rev B 58:10911
2. Nitzan A, Ratner MA (2003) Science 300:1384
3. Davis WB, Svec WA, Ratner MA, Wasielewski MR (1998) Nature 396:60
4. Schatz GC, Ratner MA (2002) Quantum mechanics in chemistry, 2 edn. Dover Publications, Mineola, New York
5. Linderberg J, Ohrn Y (2004) Propagators in quantum chemistry. Wiley, Hoboken, New Jersey
6. Pople JA, Beveridge DL (1970) Approximate molecular orbital theory. McGraw-Hill, New York
7. Jortner J (1976) J Chem Phys 64:4860
8. Marcus RA, Sutin N (1986) Biochim Biophys Acta 811:265
9. Kramers HA (1934) Physica 1:182
10. Anderson PW (1950) Phys Rev 79:350
11. Anderson PW (1959) Phys Rev 115:2
12. Nitzan A (2001) Ann Rev Phys Chem 52:681
13. Berlin YA, Burin AL, Ratner MA (2002) Chem Phys 275:61
14. Grozema FC, Berlin YA, Siebbeles LDA (2000) J Am Chem Soc 122:10903
15. Finklea HO, Hanshewm DD (1992) J Am Chem Soc 114:3173
16. Slowiski K, Chamberlain RV, Miller CJ, Majda M (1997) J Am Chem Soc 119:11910
17. Chidsey CED (1991) Science 251:919
18. Leland BA, Joran AD, Felker PM, Hopfield JJ, Zewail AH, Dervan PB (1985) J Phys Chem 89:5571
19. Oevering H, Paddon-Row MN, Heppener M, Oliver AM, Costaris E, Verhoeven JW, Hush NS (1987) J Am Chem Soc 109:3258
20. Paulson B, Pramod K, Eaton P, Closs GL, Miller JR (1993) J Phys Chem 97:13402
21. Closs GL, Miller JR (1988) Science 240:440
22. Klan P, Wagner PJ (1998) J Am Chem Soc 120:2198
23. Paddon-Row MN (1994) Acc Chem Res 27:18
24. Osuka A, Maruyama K, Mataga N, Asahi T, Yamazaki I, Tamai N (1990) J Am Chem Soc 112:4958
25. Helms A, Heiler D, McClendon G (1992) J Am Chem Soc 114:6227
26. Kim Y, Lieber CM (1989) Inorg Chem 28:3990
27. Weiss EA, Ahrens MJ, Sinks LE, Gusev AV, Ratner MA, Wasielewski MR (2004) J Am Chem Soc 126:5577
28. Helms A, Heiler D, McClendon G (1991) J Am Chem Soc 113:4325
29. Osuka A, Satoshi N, Maruyama K, Mataga N, Asahi T, Yamazaki I, Nishimura Y, Onho T, Nozaki K (1993) J Am Chem Soc 115:4577
30. Barigelletti F, Flamigni L, Balzani V, Collin J-P, Sauvage J-P, Sour A, Constable EC, Cargill Thompson AMW (1994) J Am Chem Soc 116:7692
31. Barigelletti F, Flamigni L, Guardgli M, Juris A, Beley M, Chodorowsky-Kimmes S, Collin J-P, Sauvage J-P (1996) Inorg Chem 35:136
32. Creager S, Yu CJ, Bamdad C, O'Connor S, MacLean T, Lam E, Chong Y, Olsen GT, Luo J, Gozin M, Kayyem JF (1999) J Am Chem Soc 121:1059
33. Sachs SB, Dudek SP, Hsung RP, Sita LR, Smalley JF, Newton MD, Feldberg SW, Chidsey CED (1997) J Am Chem Soc 119:10563
34. Sikes HD, Smalley JF, Dudek SP, Cook AR, Newton MD, Chidsey CED, Feldberg SW (2001) Science 291:1519

35. Benniston AC, Goulle V, Harriman A, Lehn J-M, Marczinke B (1994) J Phys Chem 98:7798
36. Osuka A, Tanabe N, Kawabata S, Speiser IS (1996) Chem Rev 96:195
37. Sachs SB, Dudek SP, Hsung RP, Sita LR, Smalley JF, New MD, Feldberg SW, Chidsey CED (1997) J Am Chem Soc 119:10563
38. Marczinke B (1994) J Phys Chem 98:7798
39. Osuka A, Tanabe N, Kawabata S, Grosshenny IV, Harriman A, Ziessel R (1995) Angew Chem, Int Ed Eng 34:1100
40. Grosshenny IV, Harriman A, Ziessel R (1995) Angew Chem, Int Ed Eng 34:2705
41. Marcus RA (1987) Chem Phys Lett 133:471
42. Ogrodnik A, Michel-Beyerle M-E (1989) Z Naturforsch 44a:763
43. Kilsa K, Kajanus J, Macpherson AN, Martensson J, Albinsson B (2001) J Am Chem Soc 123:3069
44. Lukas AS, Bushard PJ, Wasielewski MR (2002) J Phys Chem A 106:2074
45. Marcus RA (1965) J Chem Phys 43:679
46. Jortner J (1976) J Phys Chem 64:4860
47. McConnell HM (1961) J Chem Phys 35:508
48. Closs GL, Piotrowiak PJ, MacInnis JM, Fleming GR (1988) J Am Chem Soc 110:2652
49. Roest MR, Oliver AM, Paddon-Row MN, Verhoeven JW (1997) J Phys Chem A 101:4867
50. Paddon-Row MN, Oliver AM, Warman JM, Smit KJ, Haas MP, Oevering H, Verhoeven JW (1988) J Phys Chem 92:6958
51. Lopez-Castillo J-M, Filali-Mouhim A, Jay-Gerin J-P (1993) J Phys Chem 97:9266
52. Liang C, Newton MD (1992) J Phys Chem 96:2855
53. Newton MD (2000) Int J Quant Chem 77:255
54. Cave MD, Newton RJ(1997) In: Ratner MA, Jortner J (eds) Molecular electronics. Dekker, New York, pp 73-118
55. Jortner J, Bixon M, Langenbacher T, Michel-Beyerle ME (1998) Proc Natl Acad Sci USA 95:12759
56. Bixon M, Giese B, Wessely S, Langenbacher T, Michel-Beyerle ME, Jortner J (1999) Proc Natl Acad Sci USA 96:11713
57. Davis WB, Wasielewski MR, Ratner MA, Mujica V, Nitzan A (1997) J Phys Chem A 101:6158
58. Kharkats YI, Ulstrup J (1991) Chem Phys Lett 182:81
59. Skourtis SS, Mukamel S (1995) Chem Phys 197:367
60. Felts AK, Pollard WT, Friesner RA (1995) J Phys Chem 99:2929
61. Tang J (1994) Chem Phys 189:427
62. Reed MA (1992) Nanostructured systems. In: Reed MA (ed) Semiconductors and semimetals, Vol 35. Academic, Boston pp 1-7
63. Nitzan A, Jortner J, Wilkie J, Burin AL, Ratner MA (2000) J Phys Chem B 104:5661
64. Galperin M, Ratner MA, Nitzan A (2004) Nano Lett 4:1605
65. Troisi A, Ratner MA (2004) Nano Lett 4:591
66. Wang W, Lee T, Reed MA (2003) Physica E 19:117
67. Kushmerick JG, ALazorcik J, Patterson CH, Shashidhar R, Seferos DS, Bazan GC (2004) Nano Lett 4:639
68. Selzer Y, Cabassi MA, Mayer TS (2004) Nanotechnology 15:S483
69. Cave RJ, Newton MD (1996) Chem Phys Lett 249:15
70. Creutz, Newton MD, Sutin N (1994) Photochem Photobiol A: Chem 82:47
71. Kumar K, Kurnikov IV, Beratan DN, Waldeck DH, Zimmt MB (1998) J Phys Chem A 102:5529

72. Stuchebrukov, Marcus RA (1995) J Phys Chem 99:7581
73. Golub GH, Van Loan CF (1989) Matrix computations. Johns Hopkins University Press, Baltimore
74. Flannery BP, Teukolsky SA, Vetterlink WT (1988) Numerical recipes. Cambridge University Press, Cambridge, UK
75. Stuchebrukov AA (1994) Chem Phys Lett 225:55
76. Voityuk AA, Roesch N, Jortner J (2000) J Phys Chem B 104:9740
77. Newton MD (1991) Chem Rev 91:767
78. Troisi A, Orlandi G (2001) Chem Phys Lett 344:509
79. Yamashita J, Kondo J (1958) Phys Rev 109:730
80. Miller JS, Epstein AJ, Reiff WM (1988) Acc Chem Res 21:114
81. Feher, Okamura (1979) In: Chance B, Devault D, Frauenfelder H, Marcus RA, Schreiffer JR, Sutin N (eds) Tunneling conference. Academic, New York, p 729
82. Goodenough JB (1955) Phys Rev 100:564
83. Miller JS, Epstein AJ, Reiff WM (1988) Chem Rev 88:201
84. Franzen S, Shultz DA (2003) J Phys Chem A 107:4292
85. Awaga K, Maruyama Y (1989) J Chem Phys 91:2743
86. Kollmar C, Kahn O (1993) Acc Chem Res 26:259
87. Volk M, Haberle T, Feick R, Ogrodnik A, Michel-Beyerle ME (1993) J Phys Chem 97:9831
88. Werner H-J, Schulten K, Weller A (1978) Biochim Biophys Acta 502:255
89. Kobori Y, Sekiguchi S, Akiyama K, Tero-Kubota S (1999) J Phys Chem A 103:5416
90. Werner U, Kuhnle W, Staerk H (1993) J Phys Chem 97:9280
91. Weiss EA, Ratner MA, Wasielewski MR (2003) J Phys Chem A 107:3639
92. Liang C, Newton MD (1993) J Phys Chem 97:3199
93. Blomgren F, Larsson S, Nelsen S (2001) J Computat Chem 22:655
94. Troisi A, Ratner MA, Zimmt MB (2004) J Am Chem Soc 126:2215
95. Weihe H, Guedel HU (1997) J Am Chem Soc 119:6539
96. Shultz DA, Fico RM, Lee H, Kampf JW, Kirschbaum K, Pinkerton AA, Boyle PD (2003) J Am Chem Soc 125:15426
97. Goodenough JB (1963) Magnetism and the chemical bond. Interscience, New York
98. Heisenberg W (1926) Z Phys 38:411
99. Haberkorn R, Michel-Beyerle ME (1979) Biophys J 26:489
100. Hoff AJ (1986) Photochem Photobiol 43:727
101. Hoff AJ, Rademaker H, Van Grondelle R, Duysens LNM (1977) Biochim Biophys Acta 460:547
102. Michel-Beyerle ME, Scheer H, Seidlitz H, Tempus D (1979) FEBS Lett 110:129
103. Hore PJ, Hunter DA, McKie CD (1987) Chem Phys Lett 137:495
104. Blankenship RE, Schaafsma TJ, Parson WW (1977) Biochim Biophys Acta 461:297
105. Schulten K, Staerk H, Weller A, Werner H, Nickel B (1976) Z Phys Chem NF 101:371
106. Till U, Hore PJ (1997) Mol Phys 90:289
107. Weller A, Nolting F, Staerk H (1983) Chem Phys Lett 96:24
108. Weller A, Staerk H, Treichel R (1984) Faraday Discuss, Chem Soc 78:271
109. Lukas AS, Bushard PJ, Weiss EA, Wasielewski MR (2003) J Am Chem Soc 125:3921
110. Mori Y, Sakaguchi Y, Hayashi H (2000) J Phys Chem A 104:4896
111. Steiner UE, Ulrich T (1989) Chem Rev 89:51
112. Pardo E, Faus J, Julve M, Lloret F, Munoz MC, Cano J, Ottenwaelder X, Journaux Y, Carrasco R, Blay G, Fernandez I, Ruiz-Garcia R (2003) J Am Chem Soc 125:10770
113. Brunold TC, Gamelin DR, Solomon EI (2000) J Am Chem Soc 122:8511
114. McCleverty JA, Ward MD (1998) Acc Chem Res 31:842

115. Closs GL, Forbes MDE, Piotrowiak PJ (1992) J Am Chem Soc 114:3285
116. Staerk H, Busmann H-G, Kuhnle W, Treichel R (1991) J Phys Chem 95:1906
117. Shultz DA (2002) Comments Inorg Chem 23:1
118. Paddon-Row MN, Shephard MJ (2002) J Phys Chem A
119. Weiss EA, Tauber MJ, Ratner MA, Wasielewski MR (in press) J Am Chem Soc
120. Lee C, Yang W, Parr RG (1988) Phys Rev B 37:785
121. Nelsen SF, Ismagilov RF, Teki Y (1998) J Am Chem Soc 120:2200
122. Fan F, Yang J, Cai L, Price DW, Jr, Dirk SM, Kosynkin DV, Yao Y, Rawlett AM, Tour JM, Bard AJ (2002) J Am Chem Soc 124:5550
123. Zangmeister CD, Robey SW, van Zee RD, Yao Y, Tour JM (2004) J Phys Chem B 108:16187
124. Zhu X-Y (2004) J Phys Chem B 108:8778
125. Marcus RA, Sutin N (1985) Biochim Biophys Acta 811:265
126. Marcus RA (1956) J Chem Phys 24:979
127. Marcus RA (1956) J Chem Phys 24:966
128. Bixon M, Jortner J (1999) Adv Chem Phys 106:734
129. Jortner J (2003) Isr J Chem 43:169
130. Datta S (1995) Electronic transport in mesoscopic systems. Cambridge University Press, New York
131. Troisi A, Ratner MA (2003) In: Reed MA, Lee T (eds) Molecular Nanoelectronics. American Scientific Publishers, Stevenson Ranch, CA, pp 1-18
132. Seideman T, Guo H (2003) J Theor Comp Chem 2:439
133. Datta S (2005) Quantum transport: atom to transistor. Cambridge University Press, Cambridge UK
134. Grave C, Tran E, Samori P, Whitesides GM, Rampi MA (2004) Synth Met 147:11
135. Basch H, Cohen R, Ratner MA Nano Lett (in press)
136. Lee T, Wang W, Klemic JF, Zhang JJ, Su J (2004) J Phys Chem B 108:8742
137. Engelkes VB, Beebe JM, Frisbie CD (2004) J Am Chem Soc 126:14287
138. Kubatkin S, Danilov A, Hjort M, Cornil J, Bredas J-L, Sturh-Hansen N, Hedegard P, Bjornholm T (2003) Nature 425:698
139. Segal D, Nitzan A, Ratner MA, Davis WB (2000) J Phys Chem B 104:2790
140. Segal D, Nitzan A, Hanaggi P (2003) J Chem Phys 119:6840
141. Xue Y, Datta S, Ratner MA (2002) Chem Phys 281:151
142. Wingreen NS, Jauho A, Meir Y (1993) Phys Rev B 11:8487
143. Damle P, Ghosh AW, Datta S (2002) Chem Phys 281:171
144. Bihary Z, Ratner MA (2005) Phys Rev submitted
145. Nitzan A (2001) Ann Rev Phys Chem 52:681
146. Reed MA, Zhou C, Muller CJ, Burgin TP, Tour JM (1997) Science 278:252
147. Datta S, Tam W, Hong S, Riefenberger R, Henderson JI, Kubiak CP (1997) Phys Rev B 79:2530
148. McLendon G, Hake R (1992) Chem Rev 92:481
149. Winkler JR, Gray HB (1992) Chem Rev 92:369
150. Cygan MT, Dunbar TD, Arnold JJ, Bumm LA, Shedlock NF, Burgin TPI, Allara DL, Tour JM, Weiss PS (1998) J Am Chem Soc 120:2721
151. Seminario JM, Zacaias AG, Tour JM (1998) J Am Chem Soc 120:3970
152. Hsaio J-S, Krueger BP, Wagner RW, Johnson TE, Delaney JK, Mauzerall DC, Fleming GR, Lindsey JS, Bocian DF, Donohoe RJ (1996) J Am Chem Soc 118:111816
153. Wagner RW, Lindsey JS (1994) J Am Chem Soc 116:9759
154. Wagner RW, Lindsey JS, Seth J, Palaniappan V, Bocian DF (1996) J Am Chem Soc 118:3996

155. Li F, Yang SY, Ciringh Y, Seth J, Martin CH, III, Singh DL, Kim D, Birge RR, Bocian DF, Holten D, Lindsey JS (1998) J Am Chem Soc 120:10001
156. Ulman A (1996) Chem Rev 96:1533
157. Dubois LH, Nuzzo RG (1992) Ann Rev Phys Chem 43:437
158. Mann B, Kuhn HJ (1971) J Appl Phys 42:4398
159. Wong EW, Collier CP, Belohradsky M, Raymo FM, Stoddart JF, Heath JR (2000) J Am Chem Soc 122:5831
160. Collier CP, Wong EW, Belohradsky M, Raymo FM, Stoddart JF, Keukes PJ, Williams RS, Heath JR (1999) Science 285:391
161. Reed MA, Lee T (eds) (2003) Molecular Nanoelectronics. American Scientific, Stevenson Ranch, CA
162. Foley ET, Yoder NL, Guisinger NP, Hersam MC (2004) Rev Sci Instrum 75:5280
163. Xiao XY, Xu BQ, Tao NJ (2004) Nano Lett 4:267
164. Rampi MA, Schueller OJ, Whitesides GM (1998) Appl Phys Lett 72:1781
165. Slowinski K, Fong HKY, Majda M (1981) J Am Chem Soc 103:7257
166. Slowinski K, Majda M (2000) J Electroanal Chem 491:139
167. Bock CW, Trachtman M, George P (1985) Chem Phys 93:431
168. Reichert J, Ochs R, Beckmann D, Weber HB, Mayor M, Lohneysen HV (2002) Phys Rev Lett 88:176804
169. Joachim C (1999) Phys Rev B 59:12505
170. Dudek SP, Sikes HD, Chidsey CED (2001) J Am Chem Soc 123:8033
171. Davis WB, Ratner MA, Wasielewski MR (2001) J Am Chem Soc 123:7877
172. Filatov I, Larsson S (2002) Chem Phys 284:575
173. Martina N, Giacalonea F, Seguraa JL, Guldi DM (2004) Synth Met 147:57
174. Giacalonea F, Seguraa JL, Martina N, Guldi DM (2004) J Am Chem Soc 126:5340
175. Schlicke B, Belser P, De Cola L, Sabbioni E, Balzani V (1999) J Am Chem Soc 121:4207
176. Weiss EA, Ahrens MJ, Sinks LE, Ratner MA, Wasielewski MR (2004) J Am Chem Soc 126:5577
177. Q-CI (1998) Q-Chem 2.1. Q-CI, Pittsburgh, PA
178. Murphy CJ, Arkin MR, Jenkins Y, Ghatlia ND, Bossmann SH, Turro NJ, Barton JK (1993) Science 262:1025
179. Reid GD, Whittaker DJ, Day MA, Turton DA, Kayser V, Kelly JM, Beddard GS (2002) J Am Chem Soc 124:5518
180. Berlin YA, Burin AL, Ratner MA (2000) J Phys Chem A 104:443
181. Giese B (1997) Chem Res Toxicol 10:255
182. Berlin YA, Burin AL, Ratner MA (2001) J Am Chem Soc 123:260
183. Bixon M, Jortner J (2002) Chem Phys 281:393
184. Voityuk AA, Jortner J, Bixon M, Rösch N (2001) J Chem Phys 114:5614
185. Fukui K, Tanaka K, Fujitsuka M, Watanabe A, Ito O (1999) J Photochem Photobiol B: Biol 50:18
186. Kelley SO, Barton JK (1999) Science 283:375
187. Lewis FD, Wu T, Zhang Y, Letsinger RL, Greenfield SR, Wasielewski MR (1997) Science 277:673
188. Meggers E, Michel-Beyerle ME, Giese B (1998) J Am Chem Soc 120:12950
189. Lewis FD, Wu T, Liu X, Letsinger RL, Greenfield SR, Miller SE, Wasielewski MR (2000) J Am Chem Soc 122:2889
190. Wan C, Fiebig T, Schiemann O, Barton JK, Zewail AH (2000) Proc Natl Acad Sci USA 97:14052
191. Okamoto A, Tanaka K, Saito I (2003) J Am Chem Soc 125:5066
192. Xu BQ, Zhang PM, Li XL, Tao NJ (2004) Nano Lett 4:1105

193. Dekker C, Ratner MA (2001) Phys World 14:29
194. Jortner J, Bixon M, Voityuk AA, Roesch N (2002) J Phys Chem A 106:7599
195. Ten Hoeve W, Wynberg H, Havinga EE, Meijer EW (1991) J Am Chem Soc 113:5887
196. Otsubo T, Aso Y, Takimiya K (2001) Bull Chem Soc Jpn 74:1789
197. Sato M, Hiroi M (1994) Chem Lett 985
198. Bauerle P, Fischer T, Bidlingmeier B, Stabel A, Rabe JP (1995) Angew Chem, Int Ed Eng 34:303
199. Meier H, Stalmach U, Kolshorn H (1997) Acta Polym 48:379
200. Würthner F, Vollmer MS, Effenberger F, Emele P, Meyer DU, Port H, Wolf HC (1995) J Am Chem Soc 117:8090
201. Vollmer MS, Würthner F, Effenberger F, Emele P, Meyer DU, Stumpfig T, Port H, Wolf HC (1998) Chem Eur J 4:260
202. Effenberger F, Grube G (1998) Synthesis 1998:1372
203. Knorr S, Grupp A, Mehring M, Grube G, Effenberger F (1999) J Chem Phys 110:3502
204. Sato M, Fukui K, Sakamoto M, Kashiwagi S, Hiroi M (2001) Thin Solid Films 393:210
205. Otsubo T, Aso Y, Takimiya K (2002) J Mater Chem 12:2565
206. Nakamura T, Fujitsuka M, Araki Y, Ito O, Ikemoto J, Takimiya K, Aso Y, Otsubo T (2004) J Phys Chem B 108:10700
207. Hicks RG, Nodwell MB (2000) J Am Chem Soc 122:6747
208. McCullough RD (1998) Adv Mater 10:93
209. Purcell ST, Garcia N, Binh VT II, Tour JM (1994) J Am Chem Soc 116:11985
210. Tour JM, Jones L, Pearson DL, Lamba JJS, Burgin TP, Whitesides GM, Allara DL, Parikh AN, Atre SV (1995) J Am Chem Soc 117:9529
211. Wagner RW, Lindsey JS (1994) J Am Chem Soc 116:9759
212. Heilemann M, Tinnefeld P, Sanchez Mosteiro G, Garcia Parajo M, Van Hulst NF, Sauer M (2004) J Am Chem Soc 126:6514
213. Hopfield JJ (1974) Proc Natl Acad Sci USA 71:3640
214. Scholes GD, Fleming GR (2000) J Phys Chem B 104:1854
215. Donhauser ZJ, Mantooth BA, Kelly KF, Bumm LA, Monnell JD, Stapleton JJ, Price DW Jr, Rawlett AM, Allara DL, Tour JM, Weiss PS (2001) Science 292:2303
216. Giese B, Wessely S, Spormann (1999) Angew Chem, Int Ed Eng 38
217. Meggers E, Michel-Beyerle ME, Giese B (1998) J Am Chem Soc 120:12950

Top Curr Chem (2005) 257: 135–164
DOI 10.1007/b136065

Published online: 6 July 2005

The Opto-Electronic Properties of Isolated Phenylenevinylene Molecular Wires

Ferdinand C. Grozema[1,2] (✉) · Laurens D. A. Siebbeles[1] · Gerwin H. Gelinck[1,3] · John M. Warman[1]

[1]Department of Radiation Chemistry, Interfaculty Reactor Institute, Delft University of Technology, Mekelweg 15, 2629 JB Delft, The Netherlands
grozema@tnw.tudelft.nl

[2]Laboratory for Inorganic Chemistry and PCMT, DelftChemTech, Delft University of Technology, Julianalaan 136, 2628 BL Delft, The Netherlands
grozema@tnw.tudelft.nl

[3]Integrated Device Technologies, Philips Research Eindhoven, 5656 AA Eindhoven, The Netherlands

1	**Introduction**	136
2	**Exciton Formation and Decay**	137
2.1	Oligomers	138
2.2	Polymers	140
2.3	Broken-Conjugation Polymers	141
3	**Exciton Polarizability**	142
3.1	Experimental Measurements	143
3.2	Theoretical Studies of Exciton Polarizabilities	145
4	**Exciton Dissociation**	148
5	**Mobility of Electrons and Holes**	149
5.1	Experimental Measurements	149
5.2	Theoretical Studies	151
6	**Optical Absorption of PV Radical Cations**	154
6.1	Experimental Cation Spectra	154
6.2	Theoretical Studies	157
7	**Experimental Methods**	159
7.1	Materials	159
7.2	Optical Absorption and Emission	160
7.3	Flash-Photolysis Time-Resolved Microwave Conductivity (FP-TRMC)	160
7.4	Pulse-Radiolysis Time-Resolved Microwave Conductivity (PR-TRMC)	161
7.5	Pulse-Radiolysis Time-Resolved Optical Absorption (PR-TROA)	162
	References	162

Abstract The optoelectronic properties of dilute solutions of oligomeric and (broken-conjugation) polymeric phenylenevinylene chains were studied using the following techniques: optical absorption and (time-resolved) emission spectrophotometry, flash-photolysis time-resolved (real and imaginary) microwave conductivity, pulse-radiolysis time-resolved microwave conductivity, and pulse-radiolysis time-resolved optical absorption spectrophotometry. The following properties were determined: absorption and emission spectra, fluorescence quantum yields and decay times, exciton polarizabilities and dissociation probabilities, charge mobilities, and radical cation absorption spectra. The experimental results are compared with theoretical calculations of exciton polarizabilities, charge mobilities, and radical cation absorption spectra.

Keywords Molecular wire · Conjugated polymer · Phenylenevinylene · Excess polarizability · Charge carrier mobility

1 Introduction

In 1970 Hörhold and Opferman first reported that poly-phenylenevinylene (PPV) was a "moderately good" organic photoconductor [1, 2]. However, little further interest was shown in the optoelectronic properties of this π-bond-conjugated material until the discovery of the electroluminescent properties of PPV in Cambridge almost 20 years later [3]. A significant subsequent advance was the replacement of the precursor route to PPV thin-film preparation by soluble PPV derivatives which could be deposited on substrates by more generally applicable techniques such as spin-coating and drop-casting [4–8]. The availability of soluble derivatives [in particular the now "classic" poly(2-methoxy-5-(2′-ethylhexyloxy)-1,4-phenylenevinylene derivative(MEH-PPV)], made it possible in addition to highly purify such polymeric materials, an essential requirement for fluorescence studies. The ready solubility also opened up the possibility of carrying out fundamental studies of the opto-electronic properties of the isolated polymer chains [9]. Literally hundreds of papers have appeared in the subsequent decade in which a great variety of experimental and theoretical techniques have been focused on obtaining a full understanding of the photophysical properties of PPV oligomeric and polymeric chains, together with the numerous additional π-bond conjugated compounds that have been synthesized. The interested reader is directed to several review articles which deal specifically with the opto-electronic properties of isolated chains (i.e. dilute solutions) of oligomeric and polymeric phenylenevinylene (PV) derivatives [10–12].

In this paper we present a review of the experimental results that have been obtained in our group on dilute solutions of a variety of PV oligomers and polymers, some of which are shown in Fig. 1. In addition to conventional spectrophotometric measurements of absorption and emission spectra, fluorescence quantum yields and decay times, we present results on the excess polarizability and heterolytic dissociation of the relaxed S_1 excitonic state and the mobility and absorption spectra of electrons and holes on isolated

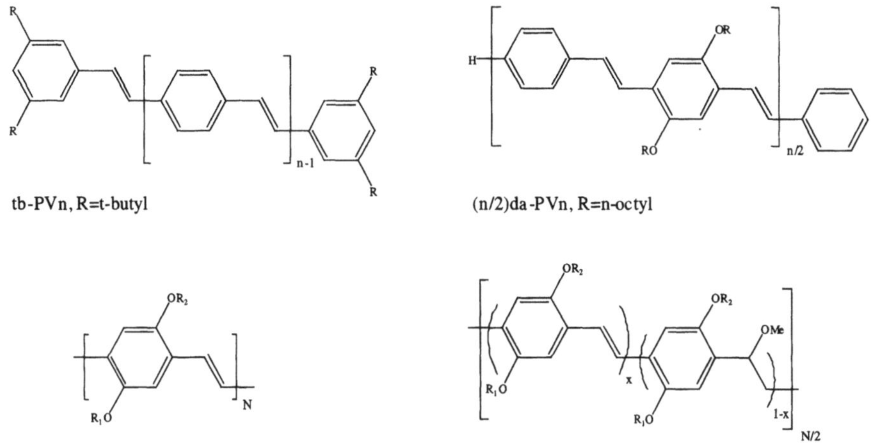

MEH-PPV: R_1=methyl, R_2=2-ethylhexyl
MDMO-PPV: R_1=methyl, R_2=3,7-dimethyloctyl
$(DMO)_2$-PPV: R_1, R_2=3,7-dimethyloctyl

MEH-PPV(x): R_1=methyl, R_2=2-ethylhexyl
MDMO-PPV(x): R_1=methyl, R_2=3,7-dimethyloctyl

Fig. 1 Molecular structures of phenylenevinylene (*PV*) oligomers and polymers investigated. Note that the numerical value of *n* in the PV*n* notation of the oligomers refers to the number of PV units and that the currently used prefix for 2-methoxy-5-3,7-dimethyloctoxy-PPV (MDMO) replaces the previous versions (dMOM and OC1OC10). *MEH-PPV* Poly(2-methoxy-5-(2′-ethylhexyloxy)-1,4-phenylenevinylene, *tb* tertiary-butyl substituted, *da* dialkoxy-substituted

PV chains, determined using flash-photolysis and pulse-radiolysis techniques with time-resolved microwave conductivity and optical absorption detection. Theoretical results on exciton polarizability and the mobility and absorption spectra of radical cation sites (holes) are also presented.

2
Exciton Formation and Decay

The formation of photo-excitations on conjugated polymer chains and their ultimate decay can be studied by the commonly used methods of absorption and emission spectrophotometry. Steady-state measurements provide information on electronic transition energies, oscillator strengths (extinction coefficients), vibrational spacings, and post-excitation structural relaxation (Stokes shift). Absolute measurement of the fluorescence quantum yield, ϕ_F, and the fluorescence decay time, τ_F, provide additional information on the rates of radiative and non-radiative transitions from the relaxed S_1 state to the ground state. In the following subsections we present spectrophotometric results obtained on dilute solutions of PV oligomers and polymers that illustrate the general trends also observed by others. Where direct comparison with results obtained by other groups could be made, good agreement was found [13–19].

2.1 Oligomers

The molecular structures of the oligomeric series of PV derivatives investigated are shown in Fig. 1. The characteristic bathochromic shift in the absorption and emission spectra with increasing chain length is illustrated for the tertiary-butyl-substituted (tb) series in Fig. 2. Also apparent is the underlying vibrational structure which is more readily seen in emission then in absorption. The first vibrational band in absorption is, in fact, only apparent as a shoulder at approximately 0.18 eV below the maximum; the same energy difference as between the 0,0 and 0,1 vibrational bands in emission.

In Table 1 are listed the energies of the absorption maximum, E_A, and the first vibrational band of the fluorescence, E_F, together with the energy difference $\Delta E_{AF} = E_A - E_F$ for all of the oligomers studied. Comparison of the results for the PV2 and PV4 oligomers in the two series illustrates the influence of dialkoxy substitution in decreasing the S_0-S_1 energy gap by about 0.25 eV. For both series, the value of ΔE_{AF} is only weakly dependent on length and equal to 0.30 eV for the longest chains. Taking into account the vibrational shoulder at 0.18 eV below E_A in the absorption spectrum mentioned above, results in a Stokes shift of 0.12 eV. Relaxation from the Franck-Condon excited state would therefore appear to be associated with only a small degree of structural rearrangement. The sharper fluorescence spectra would, however, suggest that a higher degree of backbone planarity (decrease in torsional disorder) pertains in the excited state than in the ground state.

It should be noted that the absorption spectra of oligomeric thiophene and para-phenylene chains display no resolved vibrational structure at room

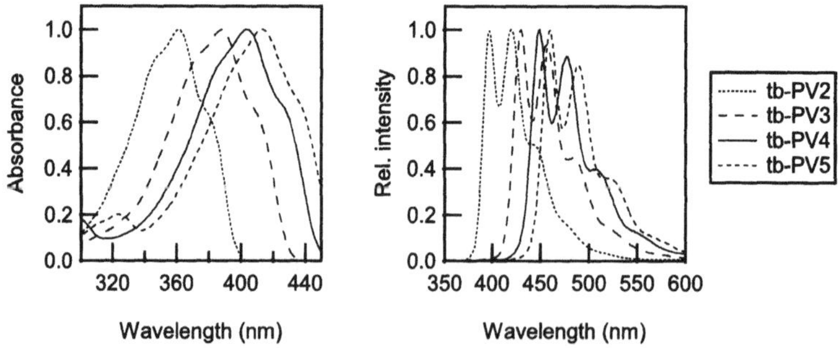

Fig. 2 Optical absorption spectra (*left*) and emission spectra (*right*) of the tertiary-butyl-substituted (*tb*) PV oligomers

Table 1 The extinction coefficient per phenylenevinylene (*PV*) unit (ε) at the first absorption maximum (E_A). The 0,0 band energy (E_F), quantum yield (ϕ_F), and decay time (τ_F) of the fluorescence, together with the calculated values of the lifetimes towards radiative (τ_R) and non-radiative (τ_{NR}) decay of the S_1 state of excess polarizability $\Delta\bar{\alpha}$

Compound	ε (10^3 M^{-1} cm^{-1})	E_A (eV)	E_F (eV)	ΔE_{AF} (eV)	τ_F (ns)	ϕ_F	τ_R (ns)	τ_{NR} (ns)	$\Delta\bar{\alpha}$ (Å^3)
tb-PV2	26.0	3.43	3.12	0.31	0.98	0.77	1.3	3.98	300
tb-PV3	19.3	3.20	2.88	0.32	0.95	0.68	1.4	2.96	530
tb-PV4	23.5	3.07	2.76	0.31	0.74	0.56	1.3	1.72	810
tb-PV5	25.8	3.00	2.70	0.30	0.67	0.57	1.2	1.52	930
1da-PV2	-	3.18	2.79	0.39	1.0	-	-		250
2da-PV4	-	2.88	2.55	0.33	0.8	-	-		660
3da-PV6	-	2.77	2.45	0.32	0.7	-	-		900
4da-PV8	-	2.72	2.42	0.30	0.61	-	-		960

temperature [20–22]. This indicates that the ground states of these molecules are subject to much greater torsional disorder than the PV oligomers. This is also evidenced by the larger values of ΔE_{AF} for oligothiophenes (0.4–0.7 eV) and oligophenylenes (0.6–0.7 eV) than for the PV oligomers, indicating that a larger rearrangement to a more coplanar structure occurs in the excited state of the former compounds. In the case of the oligophenylenes, the fluorescence spectra also display a lack of vibrational structure which may be taken to indicate a high degree of torsional disorder even in the relaxed S_1 state.

The highly emissive nature of oligo-PVs is illustrated by the fluorescence quantum yields, ϕ_F, in Table 1. A gradual decrease in ϕ_F with increasing chain length is seen to be accompanied by a decrease in the overall lifetime of the S_1 state, τ_F. Interestingly, the radiative lifetime, $\tau_R = \tau_F/\phi_F$, remains almost independent of chain length at circa 1.3 ns and the changes in ϕ_F and τ_F observed can be attributed mainly to a substantial decrease in the lifetime towards non-radiative transitions, τ_{NR}. This can be attributed to an increase in the intramolecular vibrational modes available for energy dissipation.

The extinction coefficients per monomer PV unit at the absorption maxima, ε, have been determined for the *t*-butyl-substituted derivatives and found to be insensitive to chain length with an average value of $2.4 \times 10^4\ M^{-1}\ cm^{-1}$. This value can be used to determine an approximate value of the oscillator strength per monomer unit, f, of the vertical ground state to S_1 transition using the relationship

$$f = 3.5 \times 10^{-5}\, \varepsilon \Delta \nu \qquad (1)$$

with $\Delta\nu$ the spectral half-width (FWHM) in electron volts of the first absorption band. From the average value of 0.53 eV for $\Delta\nu$, the value of f is found to be around 0.44.

2.2 Polymers

The first absorption bands of the PV polymers display no evidence of a resolved vibrational structure as found for the oligomers, although this is still apparent in emission (see Fig. 3, uppermost spectra), indicating that the torsional disorder in the ground state of the polymers is greater than in the oligomeric molecules. The values of ΔE_{AF} of 0.27 and 0.26 eV respectively are, however, both significantly lower than the value of 0.30 eV found for the longest oligomers, indicating a smaller degree of structural relaxation in S_1 for the polymers. These observations combined suggest that structural disorder in the relaxed S_1 state is also greater than for the oligomers. This is supported by the less sharply defined vibrational structure in the emission.

The absolute values of E_A and E_F are identical for the three "fully conjugated" polymers within the error of the measurements and are about 0.20 eV lower than for the dialkoxy-substituted (da)-octamer. This difference can be mainly ascribed to the fact that the polymers are dialkoxy-substituted on all phenylene rings, whereas in the octamer only alternate rings are substituted. This is supported by the observation that a decrease in E_F of 0.17 eV occurs on going from the alternately to the fully dialkoxy-substituted tetrameric PV oligomer [23]. We conclude that the S_1 to S_0 energy gap would decrease by

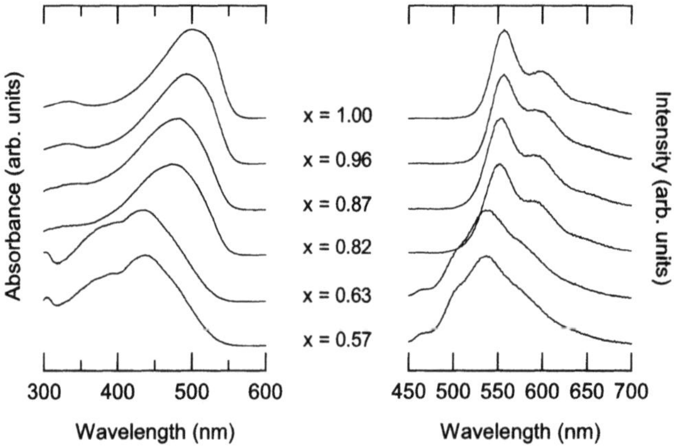

Fig. 3 Optical absorption spectra (*left*) and emission spectra (*right*) of MDMO-PPV(x) polymers with degrees of conjugation, x, from *top* to *bottom* of the vertically displaced spectra, of 1.0, 0.96, 0.87, 0.82, 0.63, and 0.57

Table 2 The energies of the first absorption band maximum (E_A) and 0,0 emission band (E_F) together with the quantum yield (ϕ_F), decay time (τ_F) of the fluorescence and the calculated values of the lifetimes towards radiative (τ_R) and non-radiative (τ_{NR}). The excess polarizability of the S_1 exciton ($\Delta\overline{\alpha}$) and the product of the charge carrier quantum yield and mobility sum ($\phi_{CS}\sum\mu$) are also listed

Compound	E_A (eV)	E_F (eV)	ΔE_{AF} (eV)	τ_F (ns)	ϕ_F	τ_R (ns)	τ_{NR} (ns)	$\Delta\overline{\alpha}$ (Å^3)	$\phi_{CS}\sum\mu$ 10^{-5} cm^2/Vs
MEH-PPV	2.47	2.20	0.27	0.25	0.20	1.6	0.31	1600	2.6
(DMO)2-PPV	2.48	2.22	0.26	0.25	0.23	1.1	0.32	1550	2.6
MDMO-PPV	2.47	2.23	0.24	0.30	0.20	1.5	0.38	1600	2.0
MDMO-PPV(x)									
$x = 0.96$	2.52	2.23	0.29	0.25	0.18	1.4	0.30	1050	1.0
$x = 0.87$	2.58	2.24	0.34	0.33	0.20	1.7	0.41	950	0.3
$x = 0.82$	2.62	2.25	0.37	0.34	0.21	1.6	0.43	975	–
$x = 0.63$	2.86	2.30	0.56	0.60	0.34	1.8	0.90	500	–
$x = 0.57$	2.84	2.30	0.54	0.60	0.36	1.7	0.93	340	–

only a few hundreths of an electron volt in going from a fully substituted octamer to the fully substituted polymer.

The trend of decreasing fluorescence yield and lifetime found for the oligomers is seen in Table 2 to be extended to the polymers, again with good agreement between the three PPVs studied. This further decrease in both parameters can also be attributed to a decrease in the non-radiative lifetime to about 0.3 ns compared with 1.5 ns for the longest tb-pentamer.

2.3 Broken-Conjugation Polymers

The effect of breaking the conjugation in the polymer (decreasing the fraction of unsaturated vinylene units, x) on the optical properties is illustrated for 2-methoxy-5-3,7-dimethyloctoxy-PPV (MDMO-PPV) in Fig. 3 and the relevant parameters are listed in Table 2. A gradual shift of the absorption maximum from 2.47 eV for the fully conjugated polymer to 2.84 eV for x close to 0.6 is apparent, and is accompanied by a considerable increase in the width of the first absorption band, particularly on the high-energy side. These effects can be explained by the increasing contribution from excitation of short conjugation length segments.

In the case of the fluorescence, no significant change in E_F is found down to $x = 0.82$ and the increase of only 0.05 eV on further reduction of x to circa 0.6 is much smaller than the overall increase of 0.37 eV in E_A. The net effect of these changes is that ΔE_{AF} increases gradually from 0.24 for the nominally fully conjugated polymer to 0.54 eV for the $x = 0.6$ broken-conjugation derivatives. The corresponding increase in Stokes shift can be explained by the occurrence of the migration of energy, subsequent to photo-excitation , from short to more highly conjugated segments of the broken-conjugation chains.

Comparison with the E_F values for the da-oligomers, taking into account the approximately 0.17 eV lower energy of the fully versus the alternately da-PV chains mentioned above, leads to the conclusion that the emission for the x=0.6 polymers arises effectively from segments with an average conjugation length of approximately six PV units. The fluorescence spectra of the MDMO-PPV(0.6) polymers are, however, seen in Fig. 3 to be much broader than those for the more conjugated polymers, indicating that emission most probably occurs from a distribution of conjugated segment lengths. In this regard, the high-energy shoulders in the MDMO-PPV(0.6) fluorescence spectra at 2.48 eV (500 nm) and 2.64 eV (470 nm) could be attributed to emission from conjugated segments only two to three PV units long.

The gradual trends found for the optical absorption and emission spectra are paralleled by changes in the fluorescence quantum yield which almost doubles, from 0.20 to 0.36 and the lifetime which increases from 0.3 to 0.6 ns. As for the oligomers discussed above, these changes in emissive properties can be ascribed mainly to a change in the non-radiative component of the decay, with τ_R remaining almost constant at 1.6 ± 0.2 ns. This is as expected if emission occurs from conjugated segments of decreasing length as x decreases.

3
Exciton Polarizability

As shown in the previous section, substantial information can be gained about the nature and decay modes of the S_1 excited state of oligomeric and polymeric PV chains from a consideration of the optical properties of dilute solutions. Questions remain, however, about the extent of delocalization of the "bound" S_1 exciton and its ability to dissociate into "free" electron and hole charge carriers.

Here, we show how the flash-photolysis time-resolved microwave conductivity technique (FP-TRMC) can provide additional information, in particular about the polarizability (extent of delocalization) and the charge separation probability of the excitonic state. The methodology depends on measuring the small changes in the real and imaginary microwave conductivity of dilute solutions which occur on flash-photo-excitation of the solute molecules. The transients measured correspond to only a few parts-per-million change in the relative dielectric constant of the solutions. The experimental methodology is reviewed briefly in Sect. 7.3. After presenting the experimental data a theoretical treatment of the exciton polarizability will be given.

3.1 Experimental Measurements

Transient changes in the imaginary component of the conductivity on photo-excitation of solutions of the tb-PV oligomers and for the MDMO-PPV(x) polymers are shown in Figs. 4 and 5 respectively. Fits to the data, taking into account the exciton lifetimes, τ_F, provide estimates of the isotropic value of the excess exciton polarizability, $\Delta\overline{\alpha} = [\overline{\alpha}(S_1) - \overline{\alpha}(S_0)]$. The values obtained for all of the compounds investigated are listed in Tables 1 and 2 and in Fig. 6.

For the two oligomeric series in Table 1, $\Delta\overline{\alpha}$ is seen to increase gradually with chain length from circa 300 Å^3 to circa 1000 Å^3. The values for the da-PV series tend to be lower by approximately 20% than for the tb-PV series. This effect of backbone substitution on the polarizability will be discussed in more detail below. Even the values of $\Delta\overline{\alpha}$ for the shortest oligomers are more than an order of magnitude larger than values found by measurements of the Stark-shift by electro-optical absorption for isolated aromatic molecules; for example, a value of 17 Å^3 for anthracene [24, 25]. Clearly the conjugated nature of the present molecules results in extensive delocalization of the exciton over the PV backbone. The results indicate that the S_1 state must be strongly coupled by an electric field (large oscillator strength) to a close-lying upper level in the singlet manifold.

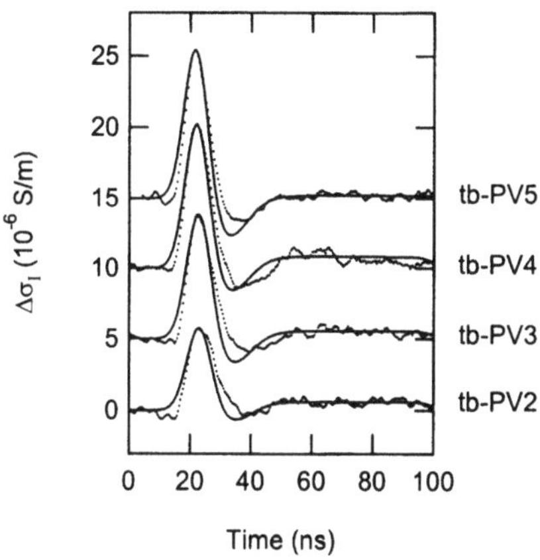

Fig. 4 *Dotted traces* are the transient changes in the imaginary component of the conductivity on flash-photolysis of OD ≈ 1 benzene solutions of the tb-PV oligomers shown on the right. The traces have been vertically displaced by 5 units for clarity. The *full curves* are calculated fits from which the exciton polarizability is determined. The noise level corresponds to a change in the relative dielectric constant of the solutions of 0.5×10^{-6}

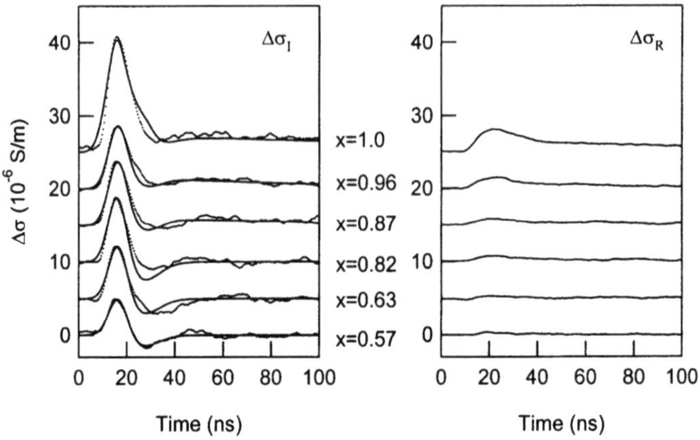

Fig. 5 The effect of broken conjugation on the transient changes in the imaginary (*left*, *dotted traces*) and real (*right*) component of the microwave conductivity on flash-photolysis of OD ≈ 1 benzene solutions of the MDMO-PPV(x) polymers. The traces have been vertically displaced for clarity. The *full curves* on the left are calculated fits from which the polarizability of the relaxed S_1 state was determined

As mentioned above, the extent of delocalization is sensitive to the nature and pattern of backbone substituents. This is illustrated by the results for the series of PV4 oligomers with different substituents shown in Fig. 6. Taking the *t*-butyl end substituents to have only a minor influence on the electronic states of the PV chain, the result for the tb-PV4 compound may be taken to be close to that expected for an unsubstituted chain. Single, dialkoxy substitution of the central phenylene ring is seen to result in a pronounced (ca 30%) reduction in the value of $\Delta\overline{\alpha}$. As will be discussed in more detail in the following theoretical section, this can be attributed to partial localization of the positive charge at the central site. On double dialkoxy substitution, the polarizability increases and on full substitution of the phenylene rings in compound 5da-PV4, the polarizability further increases to a value which is even larger than that found for the tb-PV4 derivative. In the case of 5da-PV4 all PV units are now equivalent and only end-effects will limit the magnitude of the polarizability.

In the case of the dicyano-substituted compound, CN-1da-PV4, the polarizability is seen to decrease dramatically to only 210 Å^3. In this case both the electron and hole wavefunctions are presumably concentrated in the central part of the molecule resulting in a highly localized excitonic state.

Turning to the polymers, we see that $\Delta\overline{\alpha}$ undergoes a further substantial increase to 1600 Å^3 compared with even the longest oligomers. A simple linear extrapolation of the tb-PVn results indicates that this value corresponds effectively to delocalization over eight to nine PV units. This is similar to estimates of about ten units for the exciton correlation length in PPVs made on the basis of trends in the optical properties.

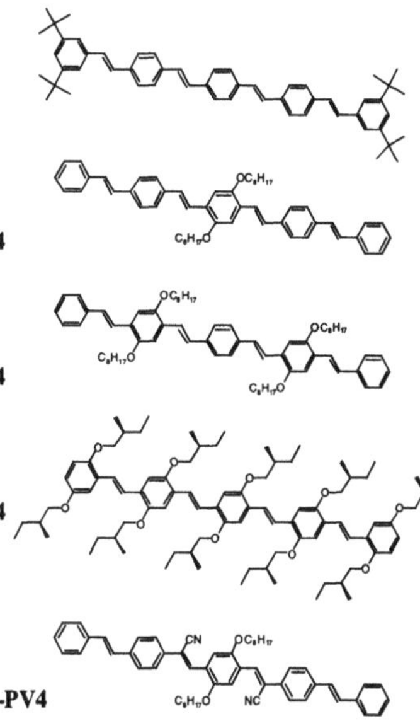

Fig. 6 The influence of substitution on the polarizability (*right column*) of the relaxed S_1 state of phenyl-capped PV tetramers. *CN* Dicyano-substituted

3.2 Theoretical Studies of Exciton Polarizabilities

Information on the properties of excited states in PVs can also be obtained by theoretical methods; however, there are only a few studies reporting calculations of excited state polarizabilities for PVs. Van der Horst et al. have calculated excited state polarizabilities for conjugated polymers by solving the Bethe-Salpeter equation to describe the excitonic state, starting from a ground state calculation using density functional theory (DFT) [26]. For PPV, the excited state polarizability was calculated to be 1000 Å^3, significantly smaller than the experimental value obtained by FP-TRMC measurements in solution (1600 Å^3) [27], but somewhat larger than the value obtained from electro-absorption measurements in a glassy tetrahydrofuran matrix (670 Å^3) [28]. It has also been shown by Van der Horst et al. that, in a sum-over-states description of the excited state polarizability, 99% or more of the polarizability of the S_1 state results from coupling to the S_2 state [26]. The same calculations for other polymers (polythiophene and poly-paraphenylene) showed that the trends in these calculations were in agreement with experimental measurements [26, 29].

In another approach it has been shown that time-dependent DFT (TDDFT) calculations combined with an external electric field can be used to calculate polarizabilities for excited states by evaluating the change in the transition energies as a function of the applied electric field [30, 31]. This way of calculating the change in polarizability on excitation is analogous to the method in which the excess polarizability is derived from electro-absorption measurements [24]. The calculated excess polarizabilities for short PVs, up to a length of four phenyl rings, are listed in Table 3. For PV2 oligomers the calculated values are in reasonable agreement with experimental results (351 Å^3 vs 420 Å^3 from electro-absorption measurements [32]). For longer chains, the calculated polarizability is considerably overestimated [30]. This is a well-known phenomenon that has also been observed in the calculation of ground state polarizabilities and has been attributed to the inadequate description of long-range electron–electron interactions in the conventional exchange-correlation functionals that were used in these calculations [33, 34].

Using computational methods it is also relatively straightforward to calculate the effect on the magnitude of the excess polarizability of variations in the molecular structure, such as changes in the nature and location of substituents. This was carried out for the PV2 oligomers shown in Fig. 7 [35]. The effect of introducing methyl groups was found to be rather small. If two methyl groups are present on the central phenyl ring (di-Me-PV2) the excess polarizability decreases from 351 Å^3 to 271 Å^3. If both outer rings are substituted with two methyl groups the excess polarizability increases slightly to 383 Å^3. However, as shown above by the experimental results the introduction of alkoxy groups can be expected to have a much larger influence on the excess polarizability.

For PV2 containing two methoxy groups on the central phenyl ring the excess polarizability is reduced by almost a factor of 4 to 91 Å^3. This shows

Table 3 Calculated values of the excitation energy (E_{exc}), the oscillator strength (f), and the excess polarizability ($\Delta\overline{\alpha}$ (calc.)) for the S_1 state together with experimental values of the last parameter ($\Delta\overline{\alpha}$ (exp.))

Compound	E_{exc} (eV)	f	$\Delta\overline{\alpha}$ (calc.) Å^3	$\Delta\overline{\alpha}$ (exp.) Å^3
PV1	-	-	47	-
PV2	2.67	1.58	351	420[a]
PV3			1283	
di-Me-PV2	2.69	1.53	271	-
tetra-Me-PV2	2.69	1.66	383	300[b]
1da-PV2	2.47	1.04	91	150[b] 340[a]
3da-PV2	2.22	1.06	808	311[a]
CN-PV2	2.52	0.52	107	-
meta-PV2	2.83	0.03	1148	-

[a] Electro-absorption measurements by Lane et al.
[b] FP-TRMC measurements by Gelinck et al.

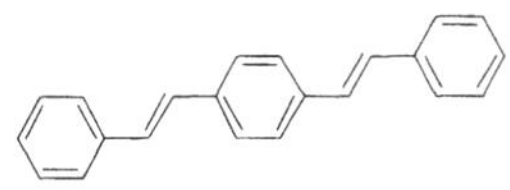

PV2

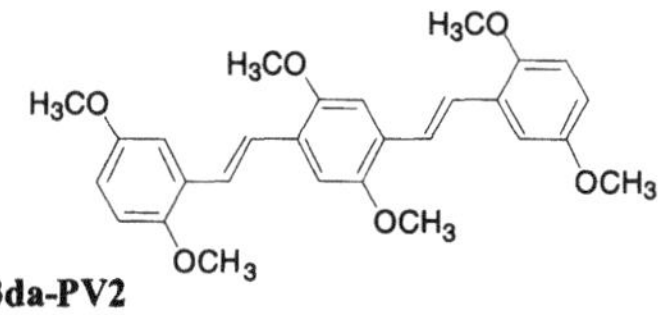

3da-PV2

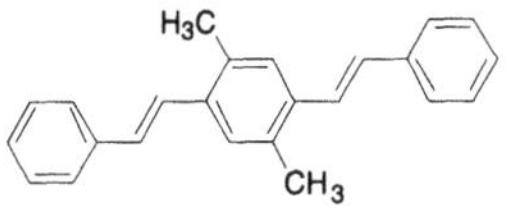

***di*-Me-PV2**

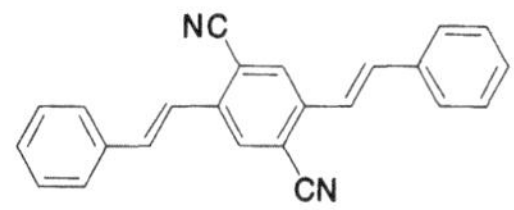

***di*-CN-PV2**

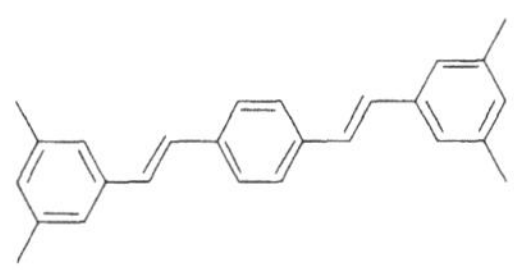

***tetra*-Me-PV2**

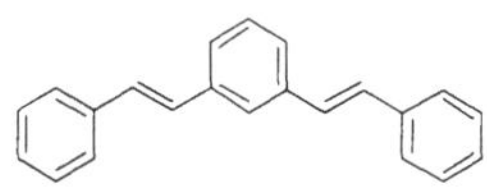

***meta*-PV2**

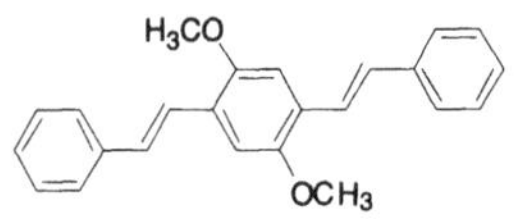

1da-PV2

Fig. 7 PV model compounds for which the excess polarizability was calculated

that the presence of the alkoxy groups strongly reduces the spatial extent of the excited state compared to the unsubstituted PV2; the exciton becomes localized on the ring where the alkoxy groups are present. These results are in qualitative agreement with experimental data from both electro-absorption measurements [32] and from FP-TRMC [23].

The effect of alkoxy substitution is further illustrated by calculations for a PV2 oligomer that bears two methoxy groups on each phenyl ring. In this case the calculated excited state polarizability increases in comparison with 1da-PV2 and becomes even larger than the value for the unsubstituted PV2. These results on the effect of the position of alkoxy side chains are in qualitative agreement with the experimental results for PV4 oligomers discussed above. It should be noted that the actual orientation of the alkoxy substituent has a significant influence on the magnitude of the excited state polarizability. The excess polarizability was calculated to be largest when the first carbon in the alkoxy chain is in the plane of the conjugated π-system.

Introduction of a cyano group was found to result in a similar lowering of the excited state polarizability, as found for methoxy groups. The large effect of cyano groups was also found experimentally for PV4 containing both methoxy and cyano groups (Fig. 6).

Variation of the PV backbone structure was found to have a dramatic effect on the magnitude of the excess polarizability. If the two styryl groups in PV2 are linked to the central phenyl ring in a meta-configuration the excess polarizability for the lowest excited state is very large (1148 $Å^3$); however, excitation to this state is only very weakly allowed from the ground state. The most strongly allowed transition was found to be due to an excitation to the third excited state. An excess polarizability of 88 $Å^3$ was found for this state, comparable to the experimental value from electro-absorption measurements (80 $Å^3$) [32].

4 Exciton Dissociation

In the previous section we presented results on the polarizability (extent of delocalization) of the relaxed S_1 excitonic state determined from measurements of the transient change in the imaginary component of the microwave conductivity on photo-excitation , $\Delta\sigma_I$. In this section we are concerned with changes in the real (dielectric loss) component, $\Delta\sigma_R$, which provides information on the formation of charge-separated dipolar or free-carrier states.

For all of the oligomeric compounds the $\Delta\sigma_R$ transients were close to the noise level of detection, indicating only a small, if any, dipolar character of the S_1 state. As shown in Fig. 5(right), this was also the case for the broken-conjugation MDMO-PPV(x) polymers up to $x = 0.82$. For higher degrees of conjugation however, a definite indication of a change in the dielectric loss on photo-excitation becomes increasingly apparent. The $\Delta\sigma_R$ transients for the DMOM-PPV(x) polymers with $x = 0.87$, 0.96 and 1.0 are shown expanded in Fig. 8 together with a transient obtained for an MEH-PPV solution under the same conditions of optical density and laser pulse intensity.

An important aspect of the $\Delta\sigma_R$ transients is that they decay over a much longer timescale than would be expected if they were attributable to the S_1 exciton itself with its sub-nanosecond lifetime. The decays are in fact quite disperse and extend into the microsecond time domain [36]. We attribute the observation of such "long-lived" dielectric loss transients to the (partial) dissociation of S_1 excitons into freely diffusing electrons and holes on the isolated polymer chains. The product of the quantum yield for charge separation, ϕ_{CS}, and the sum of the mobilities of the charge carriers, $\sum\mu$, can be calculated from the end-of-pulse value of $\Delta\sigma_R$. The values of $\phi_{CS}\sum\mu$ found are 2.6×10^{-5}, 2.0×10^{-5} and 2.6×10^{-5} cm^2/Vs for MEH-PPV, MDMO-PPV and $(MDO)_2$-PPV, respectively. Because of the random orientation of the polymer chains, $\sum\mu$ will be the effective isotropic value of the mobility, i.e. $\sum\mu_{1D}/3$ with $\sum\mu_{1D}$ the one-dimensional intrachain value. On the basis of the electron and hole mobilities on isolated MEH-PPV chains given in the next section, $\sum\mu = 0.3$ cm^2/Vs. The quantum yield for exciton dissociation into free charge carriers, on excitation at the laser wavelength used of 308 nm

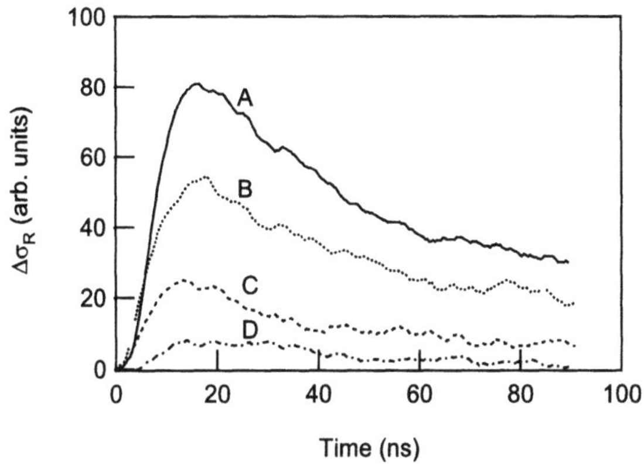

Fig. 8 The change in the real (dielectric loss) component of the microwave conductivity for benzene solutions of the following polymers: *A* MEH-PPV, *B* MDMO-PPV, *C* MDMO-PPV(0.96), *D* MDMO-PPV(0.87)

(4.0 eV), can therefore be estimated to be close to 1.0×10^{-4}. If this efficiency is the result of thermal dissociation of the relaxed exciton, then it would imply an exciton binding energy of approximately 0.25 eV. This may be taken to be a lower limit since we cannot exclude the possibility that dissociation of "hot" excitons contributes at least in part to the value of ϕ_{CS}.

An interesting complication in the study of dilute MDMO-PPV solutions was the finding that the colour changed from orange to red over a period of hours at room temperature, and on standing overnight at 5 °C a red gel phase separated from the solvent. Photo-excitation of the gel suspension resulted in $\Delta\sigma_R$ transients more than an order of magnitude larger and considerably longer lived than found for the polymer solutions [20, 36]. This is attributed to the occurrence of exciton dissociation by interchain charge transfer in the aggregated gel matrix. These results are in agreement with recent THz conductivity experiments where it was found that the upper limit for the quantum yield of photo-generation of charge carriers is 1% [37, 38].

5 Mobility of Electrons and Holes

5.1 Experimental Measurements

It goes without saying that the use of PV chains as wires in molecular electronic devices requires that they are able to transport charge. The mobility of positive and negative charges along isolated PV chains in benzene solution can be measured using the pulse-radiolysis time-resolved microwave conductivity technique (PR-TRMC) [39–41]. Charge carriers are produced in the solution by irradiation with a nanosecond pulse of 3 MeV electrons from a Van de Graaff accelerator. This produces a close-to-uniform distribution of radical cations, Bz^+, excess electrons, e^-, and excited states [42, 43]. When an oxygen-saturated benzene solution is irradiated the excess electrons and excited states react within nanoseconds with oxygen to form O_2^- and $^1\Delta_g$ O_2, respectively [44]. The Bz^+ ions are unaffected by oxygen and can diffuse to the dissolved PPV chains where they undergo charge transfer, leading to the formation of a radical cation site or "hole" on the chain. If the mobility of holes on the PPV chains is higher than that of Bz^+ in solution (1.2×10^{-3} cm^2 V^{-1} s^{-1}) the conductivity should increase initially after the irradiation pulse as the transfer of charge proceeds.

The conductivity of an oxygen saturated MEH-PPV solution after irradiation is shown as a function of time in Fig. 9. The conductivity of the solution is seen to increase over the first few hundreds of nanoseconds, which provides immediate evidence that the mobility of positive charges on the polymer chains is indeed considerably higher than the value of 1.2×10^{-3} cm^2 V^{-1} s^{-1} for the initially formed Bz^+ cations. Eventually the conductivity decays because of the recombination reaction between positively charged MEH-PPV chains and electrons that were captured by oxygen.

From a kinetic analysis of the PR-TRMC transients in which all relevant reactions between charged species are considered, the reaction rates for formation and decay, and the mobility of the charge carriers along the polymer chain can be obtained [41]. In Fig. 9, the kinetic fit obtained from such an analysis is shown as a dashed line. The mobility of the positive charge on MEH-PPV obtained from this analysis is 0.46 cm^2 V^{-1} s^{-1}.

The lower panel in Fig. 9 shows the radiation-induced conductivity of an oxygen-free MEH-PPV solution in benzene containing NH_3. The ammonia acts as a scavenger for positive charges which are thus prevented from transferring to the polymer chains. Therefore, any conductivity signal must be due to the negative charge carriers. For the electron signal, no increase in the conductivity at short times is observed. This is due to the high mobility of excess electrons in benzene (0.13 cm^2 V^{-1} s^{-1}), which leads to a very fast transfer of the electrons from benzene to the polymer. Kinetic fits in this case give a one-dimensional mobility for electrons along the MEH-PPV chains of 0.5 cm^2 V^{-1} s^{-1} [39].

It was shown in previous sections that the introduction of conjugation breaks in PPV chains has a considerable effect on optical properties and exciton delocalization. Insight into the influence of conjugation breaks on the mobility of charges has been gained by performing PR-TRMC experiments on MEH-PPV in which a controlled fraction of the vinylene groups are saturated (Fig. 1) [45]. The radiation-induced conductivity transients for $x = 1$, $x = 0.85$ and $x = 0.70$ are shown in Fig. 10. The introduction of 15% of conjugation breaks ($x = 0.85$) lowers the mobility by a factor of 10, to around 0.04 cm^2/Vs.

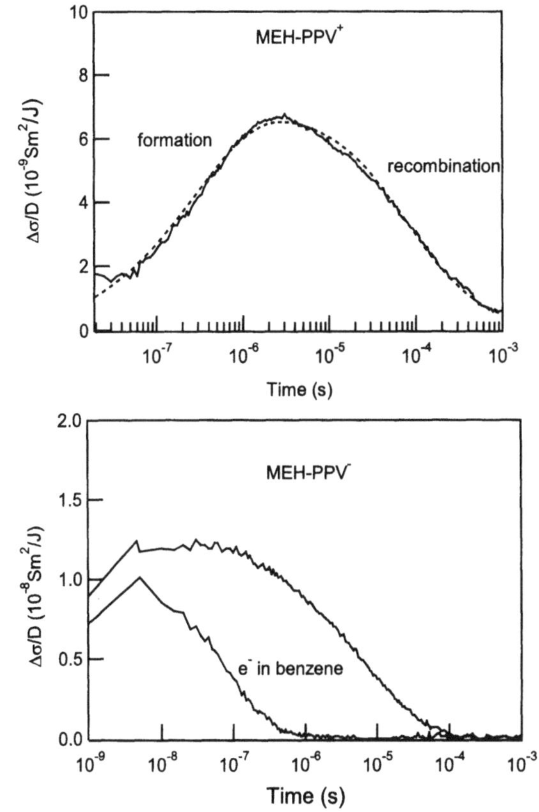

Fig. 9 *Upper panel* Transient changes in the real conductivity upon pulse-radiolysis of an oxygen saturated solution of MEH-PPV in benzene. *Lower panel* Transient changes in the real conductivity upon pulse radiolysis of pure benzene and a MEH-PPV solution in benzene containing NH_3

For $x = 0.70$ the mobility is close to the detection limit of the apparatus and was estimated to be less than 0.01 cm^2/Vs.

5.2 Theoretical Studies

Measurements of charge transport along isolated PV chains is of special interest from a theoretical point of view because isolated polymer chains are also tractable for theoretical treatments that take the inherently disordered nature of dissolved polymer chains into account. Comparison of calculated charge carrier mobilities with experimental data from PR-TRMC measurements can give valuable insights into the relationship between the molecular structure of PV chains and their conductive properties.

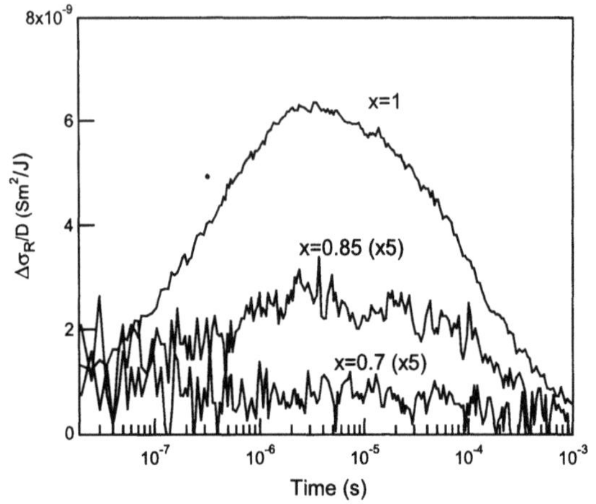

Fig. 10 Transient changes in the real conductivity on pulse radiolysis of solutions of MEH-PPV(x) in oxygen saturated benzene, with $x = 1$, 0.85 and 0.7

The mobility of charges along a PPV chain depends on the conjugation length or the degree of delocalization of the π-electron system. A planar chain conformation leads to a large conjugation length and hence to a high mobility. In order to study the effect of torsional disorder on the mobility, calculations were performed using a model based on the tight-binding approximation [46]. The polymer is modelled as a chain of sites that correspond to the monomer units. Charge transport along such a chain is described by the tight-binding Hamiltonian

$$H_{\mathrm{q}} = \sum_{n} \left[\varepsilon_n a_n^{\dagger} a_n - b(\Delta\theta_{n,n+1}) \left(a_{n+1}^{\dagger} a_n + a_n^{\dagger} a_{n+1} \right) \right] \tag{2}$$

In this equation ε_n is the energy of the charge localized at the n-th site, while $a_n^{\dagger}$ and a_n are the creation and annihilation operators for a charge at this site. The charge transfer integral, $b(\Delta\theta_{n,n+1})$, which is a measure of the electronic coupling between neighboring sites, depends on the inter-unit angle, $\Delta\theta_{n,n+1}$, between the n-th and $(n+1)$-th repeat unit in the polymer. This was calculated using quantum chemical methods [46]. The value of b is largest for a fully planar conformation of the PPV chain and becomes zero for a fully perpendicular orientation of two neighboring repeat units. This means that a regular planar chain is the optimum situation for intramolecular charge transport, as expected.

The probability of finding a point along the polymer where the inter-unit angle deviates considerably from zero is dependent on the rotational potential energy profile. Such a potential energy profile was obtained from *ab initio* electronic structure calculations. For PPV, the minimum was found for a fully planar conformation, the most favorable situation for charge transport. The inter-unit angles along a PPV chain can be sampled from a Boltzmann distribution using the potential energy profile from *ab initio* calculations [46].

The wave function, $|\Psi(t)\rangle$, of a hole is expressed as a superposition of states $|n\rangle$ located on the different sites with coefficients c_n:

$$|\Psi(t)\rangle = \sum_n c_n(t)\,|n\rangle \tag{3}$$

For hole transport the states $|n\rangle$ were taken to be the highest occupied orbitals in the neutral monomer units. At $t = 0$ a positive charge is localized on a single unit; i.e. $c_i(t = 0) = 1$ and all other coefficients are zero. The time-dependent coefficients, $c_n(t)$, were obtained by numerical integration of the first-order differential equations that follow from the substitution of the wave function in Eq. 3 into the time-dependent Schrödinger equation.

Propagation of the wave function in this way yields the mean square displacement of the charge as a function of time. From this, the frequency-dependent mobility of the charge carriers can be obtained from the mean-square displacement of the charge $\Delta^2(t)$ using the Kubo relation [47–49]

$$\mu(\omega) = \frac{-e\omega^2}{2k_{\mathrm{B}}T}\mathrm{Re}\left[\int_0^\infty \Delta^2(t)\,\exp\left(-\mathrm{i}\omega t\right)\mathrm{d}t\right] \tag{4}$$

in which e is the elementary charge, k_{B} is the Boltzmann constant, T is the temperature, ω is the radian frequency of the probing electric field, and "Re" denotes that the real part of the integral is taken.

The mobility for charges on an infinitely long polymer chain obtained using the method outlined above is plotted in Fig. 11 as a function of frequency. In the DC limit a mobility of close to 0.1 $\mathrm{cm}^2/\mathrm{Vs}$ was obtained. This is several orders of magnitude higher than the mobilities that are typically reported for DC experiments, for instance using the time-of-flight (TOF) technique, which are of the order of 10^{-5} $\mathrm{cm}^2/\mathrm{Vs}$ [50–54]. It should be noted that a direct comparison between the calculated and TOF mobility is impossible since the TOF value refers to a bulk sample where the mobility is dominated by interchain charge transport. The calculated values do, however, give an indication of the mobility that would be obtained if DC charge transport could be measured across a single PV chain.

A direct comparison of the calculated mobility with the PR-TRMC value is possible at 34 GHz. The calculated value is 12 $\mathrm{cm}^2/\mathrm{Vs}$, which is 30 times higher than the experimental hole mobility, 0.46 $\mathrm{cm}^2/\mathrm{Vs}$ [41]. The large difference between the calculated and experimental values is attributed to the finite length of the polymer chains used in the experiments and the presence of polymerization defects. Side reactions during polymerization lead to sites where single and triple bonds are present where a vinylene group should be [55]. Such defects in the polymer chain act as reflecting barriers which effectively lower the charge carrier mobility.

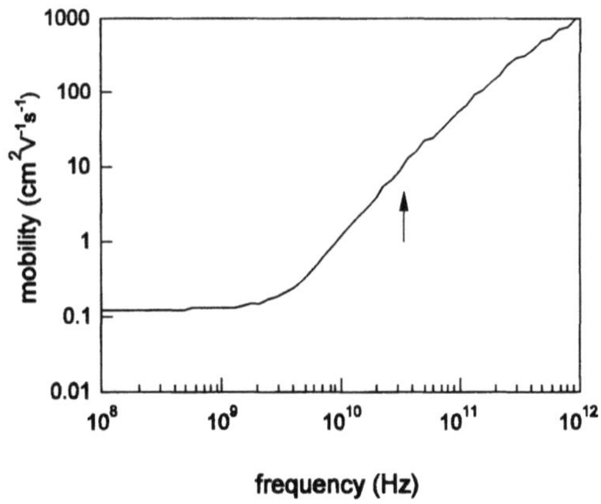

Fig. 11 Calculated real component of the mobility of positive charges on a PPV chain as a function of the frequency of the probing electric field. The *vertical arrow* indicates the microwave frequency (ca. 30 GHz) used in the experiments

In order to reproduce the experimental data, calculations were also performed for polymer chains of limited length. It was found that for a chain length of 180 units the experimental mobility at 34 GHz of 0.46 cm^2/Vs is obtained. Even though the polymer used in the experiments contains up to 1000 repeat units the effective conjugation length is considerably less because of polymerization defects that are known to occur in up to a few percent of the units. The large effect of the introduction of conjugation breaks found in these calculations is consistent with the experimental data presented in the previous section.

6 Optical Absorption of PV Radical Cations

6.1 Experimental Cation Spectra

The optical absorption spectra of PV radical cations exhibit two or three bands with energies below that of the neutral molecule [56, 57]. The position of these sub-gap transitions contains valuable information on the spatial extent of charge carriers on the PV chains. The optical absorption spectra of charged conjugated polymers are generally discussed in terms of a polaron-band model (see Fig. 12a), in which the introduction of a charge leads to the formation of two localized levels within the band-gap of the neutral polymer. In oligomers with a finite chain length the excitations are better described in terms of discrete energy levels as indicated in Fig. 12b.

Figure 13 shows the optical absorption spectra of the series of (n/2)da-PVn cations. The cations were generated by pulse radiolysis and the spectra were measured by time-resolved optical absorption, see Sect. 7.5 [57].

For 1da-PV2, two absorption bands are clearly visible. Comparison with semi-empirical quantum-chemical calculations has shown that the lowest energy band is mostly due to the excitation of an electron from the highest doubly occupied molecular orbital (H in Fig. 12b) to the singly occupied orbital (P1). The second band was found to be mostly due to excitation from the P1 level to the P2 level. For longer (n/2)da-PVn chains the lowest absorption band shifts below 0.9 eV, outside the range accessible in the experimental equipment used. The high-energy band also shifts to lower energy up to a chain length of eight repeat units. Further lengthening of the PV chain leaves the position of the second band unchanged. Interestingly, for 2da-PV4 and 3da-PV6, the second absorption band clearly exhibits a double maximum.

From electronic structure calculations it was concluded that these maxima are due to two separate close lying electronic transitions with similar oscillator strength [57]. The configuration in which the electron in the singly occupied P1 level is excited to the empty P2 level was found to contribute considerably to both these states.

In order to gain more insight into the effect of the alkoxy side chains on the PV chains additional experiments were performed for 1da-PV4 (see Fig. 6), which is dialkoxy-substituted only on the central phenyl ring [57]. The radical cation spectrum of 1da-PV4 is compared to that for 2da-PV4 in Fig. 14.

There are considerable differences between the two spectra. The high-energy band stays more or less in the same position, but its shape has com-

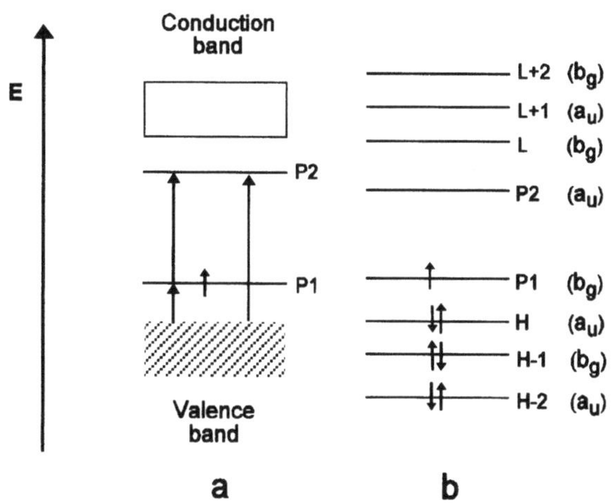

Fig. 12 Band structure model (**a**) and molecular orbital model (**b**) for sub-gap absorption features in singly charged PVs

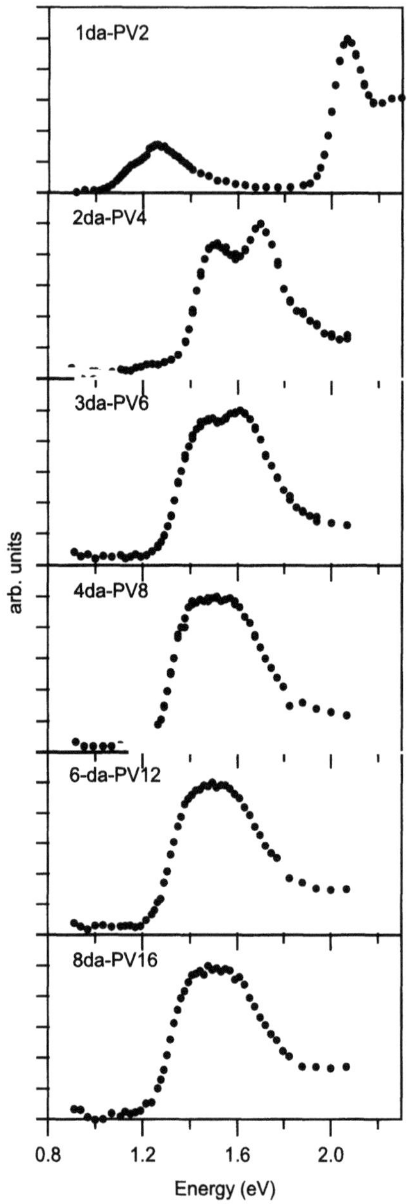

Fig. 13 Optical absorption spectra of radical cations of a series of da-PV oligomers, $(n/2)$da-PVn (see Fig. 1)

pletely changed. The onset of the low-energy absorption is clearly observed in the spectrum of 1da-PV4, while this band lies completely below 0.8 eV for 2da-PV4. The low-energy band 1da-PV4 is in fact close to that for 1da-PV2,

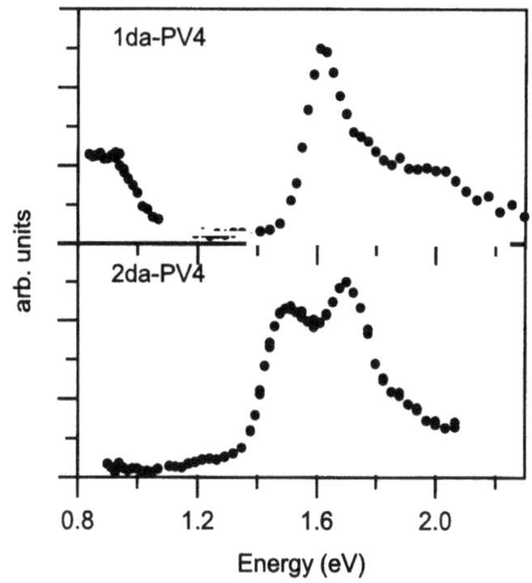

Fig. 14 Comparison of the optical absorption spectra of the radical cations of 1da-PV4 and 2da-PV4

suggesting that the spatial extent of the charge in 1da-PV4 is much smaller than in 2da-PV4.

6.2 Theoretical Studies

Quantum-chemical calculations can provide additional information on the properties of charge carriers on PV chains. Calculations have been performed for unsubstituted PVs, the fully dialkoxy-substituted $(n+1)$da-PVn and the partially substituted $(n/2)$da-PVn [57]. The geometry of all oligomers was optimized by density functional theory (DFT) calculations. Abstraction of an electron from PV oligomers leads to deformations of the geometry. These deformations are most notably present in the C – C bonds. It was found that in DFT the geometry deformations are fully delocalized over the whole chain, up to the longest chain length investigated (6da-PV12). This means that within DFT no self-localized polaron is formed. This is in contrast to earlier calculations at the Hartree-Fock (HF) level where it was found that the geometry deformation upon introduction of a positive charge comprises circa 5 PV units. These differences between HF and DFT calculations are the subject of considerable debate, and it has not yet been established which of the methods gives a more reliable description [57, 58].

Analysis of the wavefunction obtained from quantum-chemical calculations gives direct insight into the spatial extent of a charge carrier. The charge distribution of a positive charge on 1da-PV4 is compared to that for 2da-PV4 in Fig. 15. The charge distributions are based on the same wavefunction used for the calculation of the radical cation absorption spectra, see below. In 1da-PV4 the charge is localized around the central phenyl ring that is dialkoxy-substituted; more than a third of the charge is localized here. In 2da-PV4 the charge is much more evenly distributed over the chain.

The effect of the spatial extent of a charge on the radical cation absorption spectrum is most clearly seen in the position of the low- energy absorption band, as is evident from the spectra in Fig. 13. The radical cation absorption spectra of the three series of PV oligomers mentioned above have been calculated. The spectra were obtained from configuration interaction calculations based on an intermediate neglect of differential overlap (INDO) reference wave functions. The geometry was obtained from DFT calculation as mentioned above. The transition energies for the low-energy absorption band of the three oligomer series obtained from these calculations are listed in Table 4. Where available, the experimental transition energies are also given.

All calculated transition energies are in reasonable agreement with the experimental data; the maximum difference is 0.14 eV. Comparison of the transition energies for the PVn series and the $(n+1)$da-PVn series shows that dialkoxy substitution leads to a lowering of the transition energy. A similar lowering of the excitation energy upon increasing the amount of dialkoxy side chain was also observed in the absorption for the neutral spectra of PV oligomer as discussed in Sect. 2. This indicates that the oxygen atoms in the alkoxy side-chains contribute significantly to the π-electron system of the PV chain.

If not all phenyl rings are dialkoxy-substituted, interesting changes in the transition energies for the low-energy transition of the cation are observed. If

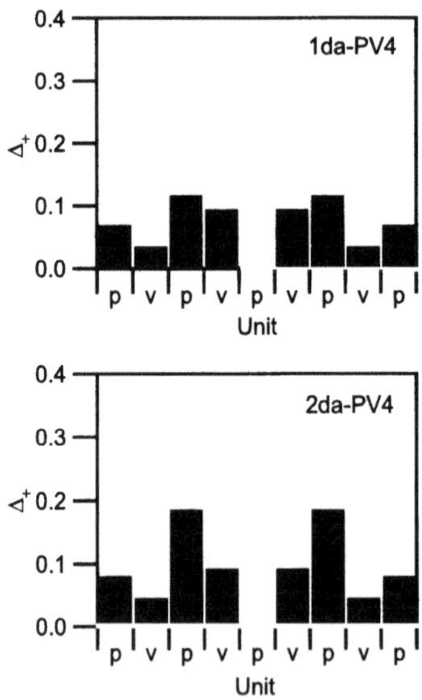

Fig. 15 Calculated charge distribution in 1da-PV4 and 2da-PV4. The letters p and v on the horizontal axis denote the positions of phenyl and vinylene groups in the chain

Table 4 Calculated transition energies for the low energy cation absorption band for the oligomer series PV*n*, (*n* + 1)da-PV*n* and (*n*/2)da-PV*n*. Experimental values are given in *parentheses*

Compound	Unsub. (eV)	(*n* + 1)da (eV)	(*n*/2)da (eV)
PV2	1.13 (1.03)	0.90 (0.76)	1.26 (1.26)
PV3	0.87 (0.80)	0.71 (0.66)	–
PV4	0.71	0.68 (0.59)	0.51
PV5	–	0.57 (0.54)	–
PV6	0.55	–	0.75

only a single phenyl ring is dialkoxy-substituted the localization of the charge at that site leads to a significant increase in the transition energy of the low-energy cation band, both in the experimental and calculated data, see Table 4. This is already apparent for PV2 where there the experimental transition energy changes from 1.03 eV to 1.26 eV upon alkoxy substitution of the central ring. The same trend was experimentally observed for 1da-PV4 and 2daPV4 in Fig. 14. An interesting case is encountered in 2da-PV4 where the degree of delocalization is actually larger than for the unsubstituted PV4. It was in fact shown that the charge distribution for PV4 has a maximum in the middle of the chain while for 2da-PV the charge on the three central phenyl rings is almost the same (see Fig. 15) [57].

7 Experimental Methods

7.1 Materials

The tb-PV, partial-da-PV, and full-da-PV oligomers were provided by the groups of professor K. Müllen (Max-Planck-Institute, Mainz), Prof. G. Hadziioannou (University of Groningen), and Prof. R.A.J. Janssen (University of Eindhoven) respectively. Samples of MEH-PPV and MDMO-PPV(x) were provided by the groups of Prof. A.B. Holmes (University of Cambridge) and Dr. E.G.J. Staring (Philips Research Laboratories). The broken conjugation versions of MEH-PPV(*x*) were provided by the group of Prof. G. Padmanaban (Indian Institute of Science, Bangalore). The molecular structures of the compounds are shown in Figs. 1 and 5.

All of the microwave conductivity measurements were carried out on solutions in Fluka "UV spectroscopic"-grade benzene. Dipolar solvents (even toluene) are unsuitable because of their large background dielectric loss at the microwave frequencies used. Dipolar solvents such as chloroform or tetrahydrofuran could be used, however, for the optical measurements.

While the PV oligomers were in general readily dissolved in benzene, polymeric PVs display a large range of solubilities. Thus, MEH-PPV solutions are permanently stable at room temperature; solutions of the branched chain dMOM- and (dMO)2-PPV polymers are stable above about 50 °C, but slowly (over a period of hours) gel at room temperature; symmetrically substituted di-*n*-alkoxy PPVs aggregate strongly and were unsuitable for the present dilute-solution investigations. The broken-conjugation polymers were all much more soluble than their "fully conjugated" analogues.

7.2 Optical Absorption and Emission

Steady-state optical absorption and emission spectra were recorded using a Kontron Uvikon-940 UV/Vis spectrophotometer and a PTI Quantamaster spectrofluorimeter respectively. Fluorescence quantum efficiencies were measured by comparing the integrated areas under the emission spectra of the solutions with standard solutions of Rhodamine 6 G in ethanol (ϕ_{FL} = 0.95) or 9,10-diphenylanthracene in cyclohexane (ϕ_{FL} = 0.90). Fluorescence decay times, τ_{FL}, were measured by flash-photolysis using 0.8 ns FWHM, 2 mJ pulses from a PRA LN1000 nitrogen laser with the emitted light detected using either an ITL TF1850 photodiode or a Photek PMT-113-UHF channel-plate photomultiplier. The estimated error in τ_{FL} was ±50 ps.

7.3 Flash-Photolysis Time-Resolved Microwave Conductivity (FP-TRMC)

In FP-TRMC studies of dilute solutions, the liquid was contained in an X-band (8.2–12.4 GHz) microwave resonant cavity with a quartz-covered grating for transmission of the light pulse. The excitation source was a Lumonics "HyperEX 400" excimer laser which delivered circa 7 ns FWHM pulses of 308 nm light with an energy density at the position of the cell of ca 5 mJ/cm^2/pulse. The pulse form and beam power were routinely monitored. The laser was used in (computer-driven) single-pulse mode with transients from a maximum of 32 pulses being averaged to improve the signal-to-noise ratio.

Transient changes in the microwave conductivity (dielectric permittivity) of the solution on photo-excitation were monitored, with nanosecond time-resolution, as changes in the microwave power reflected by the cavity. In the initial experiments measurements were made only at the resonance frequency of the cavity [59, 60]. This provides information only on changes in the real component of the conductivity (the dielectric loss) which is related to the occurrence of charge separation and the formation of mobile dipoles or freely diffusing charges.

In a later development [20, 29, 60, 61], transients were measured also at the upper and lower half-power frequencies, f^+ and f^-, of the cavity resonance. By addition and subtraction of these Transients, separate information could be gained on changes in the real and the imaginary components of the conductivity, $\Delta\sigma_R$ and $\Delta\sigma_I$. The latter provides information on the polarizability or degree of delocalization of excited states which is a parameter of great interest in the case of conjugated oligomers and polymers.

The microwave circuitry and components used were conventional and commercially available, as were the amplifiers and digitizers used for recording transients. Full details of the equipment and the data handling procedures are to be found in references [20] and [61].

7.4
Pulse-Radiolysis Time-Resolved Microwave Conductivity (PR-TRMC)

In PR-TRMC studies the solution was contained in a cell consisting of a length of Ka-band (26.5–40 GHz) rectangular waveguide terminated at one end by a metal end-plate and at the other by a Kapton or aramide window. The solution was homgeneously ionized by a pulse of 3 MeV electrons from a Van de Graaff accelerator. The incident side-wall of the cell was reduced to 0.4 mm in thickness to minimize attenuation of the electron beam. The pulses used were between 2 and 50 ns long with a beam current of circa 4 A. A number of single pulses could be averaged to improve the signal-to-noise ratio. The energy deposited per unit volume, D_V, was determined by dosimetry. From this the concentration of charge carrier pairs formed in the pulse, N_p, was calculated using the known value of the average energy per ionization event in benzene [62, 63].

The formation of mobile charge carriers was monitored as a decrease in the microwave power reflected by the cell, ΔP, using microwave circuitry and detection equipment which has been described fully elsewhere [64–66]. Single-pulse conductivity transients could be measured over a time window from nanoseconds to milliseconds using a transient digitizer with a pseudo-logarithmic time-base. The absolute value of the radiation-induced conductivity, $\Delta\sigma$, was calculated from ΔP using the known sensitivity factor for the solution-filled cell. Only information on changes in the real component of the conductivity could be determined using this technique. Fits to the conductivity transients allowed the determination of the mobility of charge carriers on the polymer chains. Additional information on the kinetics of formation and decay of the charged polymer chains could also be derived. Separate information on negatively and positively charged polymer chains could be obtained by using competitive hole or lectron-scavenging solutes. The underlying reactions and data analysis procedure have been fully described elsewhere [41].

7.5 Pulse-Radiolysis Time-Resolved Optical Absorption (PR-TROA)

In PR-TROA experiments the solution was contained in a 12.5 mm optical pathlength quartz cell and was continually refreshed during the measurements by flowing the solution through the cell from a reservoir. The solution was irradiated using a pulse of 3 MeV electrons from a Van de Graaff accelerator, as for the PR-TRMC experiments. Changes in the optical absorption of light (from a pulsed xenon arc lamp) passing through the solution were monitored as a function of wavelength using a Jobin Yvon HL300 monochromator and a photomultiplier or, for NIR detection, an InGaAs photodiode.

References

1. Hörhold HH, Opfermann J (1970) Makromol Chem 131:105
2. Hörhold HH, Opfermann J (1974) J Prakt Chem 316:750
3. Burroughes JH, Bradley DDC, Brown AR, Marks RN, MacKay K, Friend RH, Burns PL, Holmes AB (1990) Nature 347:539
4. Braun D, Heeger AJ (1991) Appl Phys Lett 58:1982
5. Murray MM, Holmes AB (2000) In: Hadziioannou G, Van Hutten PF (eds) Semiconducting polymer, chemistry, physics and engineering. Wiley-VCH, Weinheim, p 1
6. Harrison MG, Friend RH (1998) In: Müllen K, Wegner G (eds) Electronic materials: the oligomer approach. Wiley-VCH, Weinheim, p 515
7. Pope M, Swenberg CE (1999) Electronic processes in organic crystals and polymers. Oxford University Press, Oxford
8. Conwell EM (2001) In: Farchioni R, Grosso G (eds) Organic electronic materials, conjugated polymers and low molecular weight organic solids. Springer-Verlag, Berlin, p 127
9. Moses D (1992) Appl Phys Lett 60:3215
10. Samuel IDW, Rumbles G, Friend RH (1997) In: Sariciftci NS (ed) Primary photoexcitation s in conjugated polymers: molecular exciton versus semiconductor band model. World Scientific, Singapore, p 140
11. Conwell EM (1997) In: Sariciftci NS (ed) Primary Photo-excitation s in conjugated polymers: molecular exciton versus semiconductor band model. World Scientific, Singapore, p 99
12. Rothberg L (1997) In: Sariciftci NS (ed) Primary photo-excitation s in conjugated polymers: molecular exciton versus semiconductor band model. World Scientific, Singapore, p 129
13. Bredas JL, Cornil J, Beljonne D, dos Santos DA, Shuai Z (1999) Acc Chem Res 32:267
14. Nguyen T-Q, Doan V, Schwartz BJ (1999) J Chem Phys 110:4068
15. Gettinger CL, Heeger AJ, Drake JM, Pine DJ (1994) J Chem Phys 101:1673
16. Oelkrug D, Gierschner J, Egelhaaf H-J, Luer L, Tompert A, Müllen K, Stalmach U, Meier H (2001) Synth Met 121:1693
17. Gierschner J, Mack H-G, Lüer L, Oelkrug D (2002) J Chem Phys 116:8596
18. Tian B, Zerbi G, Schenk R, Mullen K (1991) J Chem Phys 95:3191
19. Mulazzi E, Ripamonti A, Wery J, Dulieu B, Lefrant S (1999) Phys Rev B 60:16519

20. Gelinck GH (1998) Excitons and polarons in luminescent conjugated polymer. Thesis, Delft University of Technology
21. Egelhaaf H-J, Oelkrug D, Gebauer W, Sokolowski M, Umbach E, Fischer T, Bäuerle P (1998) Opt Mater 9:59
22. Nijegorodov NI, Downey WS, Danailov MB (2000) Spectrochim Acta A 56:783
23. Candeias LP, Gelinck GH, Piet JJ, Piris J, Wegewijs BR, Peeters E, Wildeman J, Hadziioannou G, Müllen K (2001) Synth Met 119:339
24. Liptay W, Walz G, Baumann W, Schlosser H-J, Deckers H, Detzer N (1971) Z Naturforsch 26a:2020
25. Mathies R, Albrecht AC (1974) J Chem Phys 60:2500
26. van der Horst J-W, Bobbert PA, de Jong PHL, Michels MAJ, Siebbeles LDA, Warman JM, Gelinck GH, Brocks G (2001) Chem Phys Let 334:303
27. Gelinck GH, Piet JJ, Wegewijs BR, Müllen K, Wildeman J, Hadziioannou G, Warman JM (2000) Phys Rev B 62:1489
28. Wachsmann-Hogiu S, Peteanu LA, Liu LA, Yaron DJ, Wildeman J (2003) J Phys Chem B 107:5133
29. Warman JM, Gelinck GH, Piet JJ, Suykerbuyk JWA, de Haas MP, Langeveld-Voss BMW, Janssen RAJ, Hwang D-H, Holmes AB, Remmers M, Neher D, Müllen K, Bäuerle P (1997) Proc SPIE 3145:142
30. Grozema FC, Telesca R, Jonkman HT, Siebbeles LDA, Snijders JG (2001) J Chem Phys 115:10014
31. Grozema FC (2003) Opto-electronic properties of conjugated molecular Wires. Thesis, Delft University of Technology
32. Lane PA, Mellor H, Martin SJ, Hagler TW, Bleyer A, Bradley DDC (2000) Chem Phys 257:41
33. Van Faassen M, de Boeij PL, Van Leeuwen R, Berger JA, Snijders JG (2002) Phys Rev Lett 88:186401
34. Van Faassen M, de Boeij PL, Van Leeuwen R, Berger JA, Snijders JG (2003) J Chem Phys 118:1044
35. Grozema FC, Telesca R, Snijders JG, Siebbeles LDA (2003) J Chem Phys 118:9441
36. Gelinck GH, Warman JM, Staring EGJ (1996) J Phys Chem 100:5485
37. Hendry E, Schins JM, Candeias LP, Siebbeles LDA, Bonn M (2004) Phys Rev Lett 92:196601
38. Hendry E, Koeberg M, Schins JM, Siebbeles LDA, Bonn M (in press) Phys Rev B
39. Hoofman RJOM, de Haas MP, Siebbeles LDA, Warman JM (1998) Nature 392:54
40. Grozema FC, Siebbeles LDA, Warman JM, Seki S, Tagawa S, Scherf U (2002) Adv Mater 14:228
41. Grozema FC, Hoofman RJOM, Candeias LP, de Haas MP, Warman JM, Siebbeles LDA (2003) J Phys Chem A 107:5976
42. Cooper R, Thomas JK (1968) J Chem Phys 48:5097
43. Thomas JK, Mani I (1969) J Chem Phys 51:1834
44. Gorman AA, Lovering G, Rodgers MAJ (1978) J Am Chem Soc 100:4527
45. Candeias LP, Grozema FC, Padmanaban G, Ramakrishnan S, Siebbeles LDA, Warman JM (2002) J Phys Chem B 107:1554
46. Grozema FC, van Duijnen PT, Berlin YA, Ratner MA, Siebbeles LDA (2002) J Phys Chem B 106:7791
47. Dyre JC, Schrøder TB (2000) Rev Mod Phys 72:873
48. Kubo R (1957) J Phys Soc Jpn 12:570
49. Scher H, Lax M (1973) Phys Rev B 7:4491

50. Gailberger M, Bässler H (1991) Phys Rev B 44:8643
51. Meyer H, Haarer D, Naarman H, Hörhold HH (1995) Phys Rev B 52:2587
52. Lebedev E, Dittrich T, Petrova-Koch V, Karg S, Brütting W (1997) Appl Phys Lett 71:2686
53. Inigo AR, Chiu HC, Fann W, Huang YS, Jeng US, Lin TL, Hsu CH, Peng KY, Chen SA (2004) Phys Rev B 69:075201
54. Mozer AJ, Denk P, Scharber MC, Neugebauer H, Sariciftci NS, Wagner P, Lutsen L, Vanderzande D (2004) J Phys Chem B 108:5235
55. Becker H, Spreitzer H, Ibrom K, Kreuder W (1999) Macromolecules 32:4925
56. Cornil J, Beljonne D, Bredas JL (1995) J Chem Phys 103:834
57. Grozema FC, Candeias LP, Swart M, van Duijnen PT, Wildeman J, Hadziioannou G, Siebbeles LDA, Warman JM (2002) J Chem Phys 117:11366
58. Geskin VM, Dkhissi A, Brédas J-L (2002) Int J Quant Chem 91:350
59. de Haas MP, Warman JM (1982) Chem Phys 73:35
60. Schuddeboom W (1994) Photophysical properties of opto-electronic molecules studied by time-resolved microwave conductivity. Thesis, Delft University of Technology
61. Piet JJ (2001) Excitonic interactions in multichromophoric arrays. Thesis, Delft University of Technology
62. Gee N, Freeman GR (1992) Can J Chem 70:1618
63. Schmidt WF, Allen AO (1968) J Phys Chem 72:3730
64. Infelta PR, de Haas MP, Warman JM (1977) Radiat Phys Chem 10:353
65. Warman JM, De Haas MP (1991) In: Tabata Y (ed) Pulse radiolysis. CRC, Boston p 101
66. Warman JM (1982) In: Baxendale JH, Busi F (eds) The study of fast processes and transient species by electron pulse radiolysis. Reidel, Dordrecht p 129

Author Index Volumes 251–257

Author Index Vols. 26–50 see Vol. 50
Author Index Vols. 51–100 see Vol. 100
Author Index Vols. 101–150 see Vol. 150
Author Index Vols. 151–200 see Vol. 200
Author Index Vols. 201–250 see Vol. 250

The volume numbers are printed in italics

Alberto R (2005) New Organometallic Technetium Complexes for Radiopharmaceutical Imaging. *252*: 1–44
Anderson CJ, see Li WP (2005) *252*: 179–192
Anslyn EV, see Houk RJT (2005) *255*: 199–229
Araki K, Yoshikawa I (2005) Nucleobase-Containing Gelators. *256*: 133–165
Armitage BA (2005) Cyanine Dye–DNA Interactions: Intercalation, Groove Binding and Aggregation. *253*: 55–76
Arya DP (2005) Aminoglycoside–Nucleic Acid Interactions: The Case for Neomycin. *253*: 149–178

Bailly C, see Dias N (2005) *253*: 89–108
Barbieri CM, see Pilch DS (2005) *253*: 179–204
Bayly SR, see Beer PD (2005) *255*: 125–162
Beer PD, Bayly SR (2005) Anion Sensing by Metal-Based Receptors. *255*: 125–162
Boschi A, Duatti A, Uccelli L (2005) Development of Technetium-99m and Rhenium-188 Radiopharmaceuticals Containing a Terminal Metal–Nitrido Multiple Bond for Diagnosis and Therapy. *252*: 85–115
Braga D, D'Addario D, Giaffreda SL, Maini L, Polito M, Grepioni F (2005) Intra-Solid and Inter-Solid Reactions of Molecular Crystals: a Green Route to Crystal Engineering. *254*: 71–94
Brizard A, Oda R, Huc I (2005) Chirality Effects in Self-assembled Fibrillar Networks. *256*: 167–218

Chaires JB (2005) Structural Selectivity of Drug-Nucleic Acid Interactions Probed by Competition Dialysis. *253*: 33–53
Chiorboli C, Indelli MT, Scandola F (2005) Photoinduced Electron/Energy Transfer Across Molecular Bridges in Binuclear Metal Complexes. *257*: 63–102
Collyer SD, see Davis F (2005) *255*: 97–124
Correia JDG, see Santos I (2005) *252*: 45–84

D'Addario D, see Braga D (2005) *254*: 71–94
Davis F, Collyer SD, Higson SPJ (2005) The Construction and Operation of Anion Sensors: Current Status and Future Perspectives. *255*: 97–124
Dervan PB, Poulin-Kerstien AT, Fechter EJ, Edelson BS (2005) Regulation of Gene Expression by Synthetic DNA-Binding Ligands. *253*: 1–31

Dias N, Vezin H, Lansiaux A, Bailly C (2005) Topoisomerase Inhibitors of Marine Origin and Their Potential Use as Anticancer Agents. *253*: 89–108
Duatti A, see Boschi A (2005) *252*: 85–115

Edelson BS, see Dervan PB (2005) *253*: 1–31
Edwards DS, see Liu S (2005) *252*: 193–216
Escudé C, Sun J-S (2005) DNA Major Groove Binders: Triple Helix-Forming Oligonucleotides, Triple Helix-Specific DNA Ligands and Cleaving Agents. *253*: 109–148

Fages F, Vögtle F, Žinić M (2005) Systematic Design of Amide- and Urea-Type Gelators with Tailored Properties. *256*: 77–131
Fages F, see Žinić M (2005) *256*: 39–76
Fechter EJ, see Dervan PB (2005) *253*: 1–31
Fujiwara S-i, Kambe N (2005) Thio-, Seleno-, and Telluro-Carboxylic Acid Esters. *251*: 87–140

Gelinck GH, see Grozema FC (2005) *257*: 135–164
Giaffreda SL, see Braga D (2005) *254*: 71–94
Grepioni F, see Braga D (2005) *254*: 71–94
Grozema FC, Siebbeles LDA, Gelinck GH, Warman JM (2005) The Opto-Electronic Properties of Isolated Phenylenevinylene Molecular Wires. *257*: 135–164

Higson SPJ, see Davis F (2005) *255*: 97–124
Hirst AR, Smith DK (2005) Dendritic Gelators. *256*: 237–273
Houk RJT, Tobey SL, Anslyn EV (2005) Abiotic Guanidinium Receptors for Anion Molecular Recognition and Sensing. *255*: 199–229
Huc I, see Brizard A (2005) *256*: 167–218

Indelli MT, see Chiorboli C (2005) *257*: 63–102
Ishii A, Nakayama J (2005) Carbodithioic Acid Esters. *251*: 181–225
Ishii A, Nakayama J (2005) Carboselenothioic and Carbodiselenoic Acid Derivatives and Related Compounds. *251*: 227–246

James DK, Tour JM (2005) Molecular Wires. *257*: 33–62
Jones W, see Trask AV (2005) *254*: 41–70

Kambe N, see Fujiwara S-i (2005) *251*: 87–140
Kano N, Kawashima T (2005) Dithiocarboxylic Acid Salts of Group 1–17 Elements (Except for Carbon). *251*: 141–180
Kato S, Niyomura O (2005) Group 1–17 Element (Except Carbon) Derivatives of Thio-, Seleno- and Telluro-Carboxylic Acids. *251*: 19–85
Kato S, see Niyomura O (2005) *251*: 1–12
Kato T, Mizoshita N, Moriyama M, Kitamura T (2005) Gelation of Liquid Crystals with Self-Assembled Fibers. *256*: 219–236
Kaul M, see Pilch DS (2005) *253*: 179–204
Kaupp G (2005) Organic Solid-State Reactions with 100% Yield. *254*: 95–183
Kawashima T, see Kano N (2005) *251*: 141–180
Kitamura T, see Kato T (2005) *256*: 219–236
Komatsu K (2005) The Mechanochemical Solid-State Reaction of Fullerenes. *254*: 185–206

Lansiaux A, see Dias N (2005) *253*: 89–108
Lhoták P (2005) Anion Receptors Based on Calixarenes. *255*: 65–95
Li WP, Meyer LA, Anderson CJ (2005) Radiopharmaceuticals for Positron Emission Tomography Imaging of Somatostatin Receptor Positive Tumors. *252*: 179–192
Liu S (2005) 6-Hydrazinonicotinamide Derivatives as Bifunctional Coupling Agents for ^{99m}Tc-Labeling of Small Biomolecules. *252*: 117–153
Liu S, Robinson SP, Edwards DS (2005) Radiolabeled Integrin $\alpha_v\beta_3$ Antagonists as Radiopharmaceuticals for Tumor Radiotherapy. *252*: 193–216
Liu XY (2005) Gelation with Small Molecules: from Formation Mechanism to Nanostructure Architecture. *256*: 1–37

Maini L, see Braga D (2005) *254*: 71–94
Matsumoto A (2005) Reactions of 1,3-Diene Compounds in the Crystalline State. *254*: 263–305
Meyer LA, see Li WP (2005) *252*: 179–192
Mizoshita N, see Kato T (2005) *256*: 219–236
Moriyama M, see Kato T (2005) *256*: 219–236
Murai T (2005) Thio-, Seleno-, Telluro-Amides. *251*: 247–272

Nakayama J, see Ishii A (2005) *251*: 181–225
Nakayama J, see Ishii A (2005) *251*: 227–246
Niyomura O, Kato S (2005) Chalcogenocarboxylic Acids. *251*: 1–12
Niyomura O, see Kato S (2005) *251*: 19–85

Oda R, see Brizard A (2005) *256*: 167–218

Paulo A, see Santos I (2005) *252*: 45–84
Pilch DS, Kaul M, Barbieri CM (2005) Ribosomal RNA Recognition by Aminoglycoside Antibiotics. *253*: 179–204
Piwnica-Worms D, see Sharma V (2005) *252*: 155–178
Polito M, see Braga D (2005) *254*: 71–94
Poulin-Kerstien AT, see Dervan PB (2005) *253*: 1–31

Ratner MA, see Weiss EA (2005) *257*: 103–133
Robinson SP, see Liu S (2005) *252*: 193–216

Sakamoto M (2005) Photochemical Aspects of Thiocarbonyl Compounds in the Solid-State. *254*: 207–232
Santos I, Paulo A, Correia JDG (2005) Rhenium and Technetium Complexes Anchored by Phosphines and Scorpionates for Radiopharmaceutical Applications. *252*: 45–84
Scandola F, see Chiorboli C (2005) *257*: 63–102
Scheffer JR, Xia W (2005) Asymmetric Induction in Organic Photochemistry via the Solid-State Ionic Chiral Auxiliary Approach. *254*: 233–262
Schmidtchen FP (2005) Artificial Host Molecules for the Sensing of Anions. *255*: 1–29 Author Index Volumes 251–255
Sharma V, Piwnica-Worms D (2005) Monitoring Multidrug Resistance P-Glycoprotein Drug Transport Activity with Single-Photon-Emission Computed Tomography and Positron Emission Tomography Radiopharmaceuticals. *252*: 155–178
Siebbeles LDA, see Grozema FC (2005) *257*: 135–164
Smith DK, see Hirst AR (2005) *256*: 237–273

Stibor I, Zlatušková P (2005) Chiral Recognition of Anions. *255*: 31–63
Suksai C, Tuntulani T (2005) Chromogenetic Anion Sensors. *255*: 163–198
Sun J-S, see Escudé C (2005) *253*: 109–148

Tobey SL, see Houk RJT (2005) *255*: 199–229
Toda F (2005) Thermal and Photochemical Reactions in the Solid-State. *254*: 1–40
Tour JM, see James DK (2005) *257*: 33–62
Trask AV, Jones W (2005) Crystal Engineering of Organic Cocrystals by the Solid-State Grinding Approach. *254*: 41–70
Tuntulani T, see Suksai C (2005) *255*: 163–198

Uccelli L, see Boschi A (2005) *252*: 85–115

Vezin H, see Dias N (2005) *253*: 89–108
Vögtle F, see Fages F (2005) *256*: 77–131
Vögtle M, see Žinić M (2005) *256*: 39–76

Warman JM, see Grozema FC (2005) *257*: 135–164
Wasielewski MR, see Weiss EA (2005) *257*: 103–133
Weiss EA, Wasielewski MR, Ratner MA (2005) Molecules as Wires: Molecule-Assisted Movement of Charge and Energy. *257*: 103–133
Williams LD (2005) Between Objectivity and Whim: Nucleic Acid Structural Biology. *253*: 77–88
Wong KM-C, see Yam VW-W (2005) *257*: 1–32

Xia W, see Scheffer JR (2005) *254*: 233–262

Yam VW-W, Wong KM-C (2005) Luminescent Molecular Rods – Transition-Metal Alkynyl Complexes. *257*: 1–32
Yoshikawa I, see Araki K (2005) *256*: 133–165

Žinić M, see Fages F (2005) *256*: 77–131
Žinić M, Vögtle F, Fages F (2005) Cholesterol-Based Gelators. *256*: 39–76
Zlatušková P, see Stibor I (2005) *255*: 31–63

Subject Index

Acetyl (4-ethynylthiophenol) 38
Acetylene oligomers 49
AFM, molecular wires 57
Alkanethiolates 39
Alkynyl transition-metal complexes 1
Alkynylplatinum(II) phosphines 8
Alligator clips, Au surfaces 37, 55
Aminocarbene 7
Au wires 56
Azy-aromatic bridges 88

Beta parameter 106
bqpy 97
Broken-conjugation polymers 141
Butadiynyl 5

Carbon nanotubes 33, 50, 56
Cation spectra 154
Charge carrier mobility 135
Charge hopping 67, 70
Charge injection 116
Charge transfer 63
- mechanism 106
- sequential 107
Chemosensors 1
Chromophores, MLCT 75
Conductance, molecular wires 33
Conductance calculations 117
Copper(I) alkynyl complex 18
Coulombic mechanism 72

Decay 137
Density functional theory 157
Dexter-type mechanism 73
DFT calculations 145
Dielectric permittivity 160
Diphenylacetylide 19
Dipole-dipole term 73
DNA 123
Donor/acceptor 63

Electroluminescence 2
Electron transfer 63, 67, 103
Electron tunneling time 109
Electronic coupling 103, 109
Electronics, molecular 34
Energy matching 116
Energy transfer 63, 65, 71, 103
Energy transport 104
Ethynylpyridine 16
Exchange energy transfer 72
Exciton dissociation 148
Exciton formation 137
Exciton polarizabilties 143, 145

FCWD 66
FP-TRMC 142, 160
Franck-Condon factor 66, 72

Gold(I) alkynyl complex 21
Gold(I) phosphine complex, ethynyl-bridged 23
Greens function 118

Hamiltonians 105
- tight-binding 104
Hartree-Fock 157
Hole transfer 65
Hopping 67, 70, 108
- mechanism 106
Human serum albumin 13

INDO 158
Iodomethane 38
Iron(II) alkynyl complex 26

Ladder oligomers, aromatic, conductivity 46
Landauer formula 117
Ligand-to-ligand charge transfer 12

Lithography 35
LLCT 12
Luminescence, transition-metal complexes 1

Magnetic exchange coupling 111
Magnetic field effect 113
Marcus theory 66
McConnell model 107
MDMO 137
MDMO-PPV 141, 149
MEH-PPV 135
Mercury 55
Metal-to-ligand charge transfer 2
Metal-to-metal electron transfer 92
MLCT 2, 84
- chromophores 75
Molecular bridges, photoinduced electron/energy transfer 63
Molecular wires 33, 36, 63, 103
- conductance 55
- organometallic 51
Moore's law 35
Mulliken-Hush theory, Generalized (GMH) 110

Nanowires 36
$NbCl_4$ 36
4-Nitrophenylacetylide 19

OLEDs 2, 36
Oligo(benzoanthracene) 46
Oligo(pentaacetylene), encapsulated 49
Oligo(1,4-phenylene ethynylene)s 33, 38
Oligo(1,4-phenylene vinylene)s 33, 43, 120
Oligophenylenes 47, 121, 139
Oligo(quinoxaline) 46
Oligo(2,5-thiophene ethynylene)s 33, 37
Oligothiophenes 125, 139
Optoelectronics, molecular wires 36
Osmium(II) 84

Phenylenevinylene (PV) 135
Phosphines, alkynylplatinum(II) 8
Phosphinogold(I) alkynyl complex 22
Phosphorescence 1
Photoelectron spectroscopy 117
Photoexitation 2
Photoinduced electron transfer 13
Photoluminescence 48
Photonic wires 126
Platinum(II) alkynyl complex, cyclometalated 17
Platinum(II) phosphines 8
Platinum(II) polypyridyl complexes 12
Platinum(II) terpyridyl complexes 13
Polarizability, excess 135
- exciton 142
Poly(9,9-diarylfluorene)s 48
Poly(phenylene vinylene)s 36
Polyindenofluorenes 48
Polyphenylenes 33, 47, 78
- bridges 63
Polypyridines 65
- platinum(II) 12
- rhenium(I) 4
- ruthenium(II) 6
Polypyridyl ruthenium(II) 6
Polyquinoxaline bridges 63, 88
Poly-ynes 1
Porphyrin-diaryl-ethyne 51
Porphyrins, metallacycles 53
- molecular wires 51
PPV 135
PR-TRMC 150, 161
PR-TROA 162
PV radical cations 154

Radical pair intersystem crossing 113
Rhenium(I) monoynyl/diynyl 4
Rhenium(I) triynyl 5
Ruthenium(II) 4, 78
Ruthenium(III) 78

SAMs 106
- conductance 55
Sequential charge transfer 107
Silver(I) alkynyl complex 18
Stokes shift 137
Superexchange 67, 68, 103, 106
SWNT 56

tatpp ligand 95
Testbeds 55
Tight-binding Hamiltonian 104
tpphz complexes 89

Wires 33
- photonic 126

Zeolites 49

Printing: Krips bv, Meppel
Binding: Stürtz, Würzburg

MIX
Papier aus verantwortungsvollen Quellen
Paper from responsible sources
FSC® C105338

If you have any concerns about our products,
you can contact us on
ProductSafety@springernature.com

In case Publisher is established outside the EU,
the EU authorized representative is:
Springer Nature Customer Service Center GmbH
Europaplatz 3, 69115 Heidelberg, Germany

Printed by Libri Plureos GmbH
in Hamburg, Germany